Other titles in the Breakthrough series:

Breakthrough: Women in Religion

Breakthrough: Women in Politics

Breakthrough: Women in Writing

Breakthrough: Women in Archaeology

Breakthrough: Women in Television

Breakthrough:
Women in Aviation

Breakthrough:
Women in Aviation

Elizabeth Simpson Smith

Foreword by Frank Borman

Walker and Company
New York

Library of Congress Cataloging in Publication Data
Smith, Elizabeth Simpson.
  Breakthrough, women in aviation.

  1. Women in aeronautics.   2. Aeronautics—
United States—Biography.   I. Title.
TL539.S6   1981   629.13′0092′2   81-7414
ISBN 0-8027-6433-9   AACR2

FIRST PUBLISHED IN THE UNITED STATES OF AMERICA IN 1981 BY THE
WALKER PUBLISHING COMPANY, INC.

PUBLISHED SIMULTANEOUSLY IN CANADA BY JOHN WILEY & SONS
CANADA, LIMITED, REXDALE, ONTARIO.

ISBN: 0-8027-6433-9

LIBRARY OF CONGRESS CATALOG CARD NUMBER: 81-50736

PRINTED IN THE UNITED STATES OF AMERICA

10   9   8   7   6   5   4   3   2   1

For Nora,
whose flights are
confined to fantasy

ACKNOWLEDGMENTS

THE AUTHOR WISHES to express sincere gratitude to the following who have assisted in research or in the preparation of this manuscript: Gerrie T. Cook, Janet Morris, Bob Adams, Bernard Groseclose, Leslie Mahaffey, Dr. Nick Komans and Bob Adams of the Federal Aviation Administration; Capt. Sam Roberts, Lt. Col. Nick Apple, 1Lt. James Pearson, Capt. Pat Osborn, Maj. Paul Kahl, Sr., Lt. Miles Wiley and Janice King of the U.S. Air Force; Dick Martin and Lee Rogers of Lockheed-Georgia Co.; Bill Davis and Jim Mummah of United Airlines; John E. Roberts of Zantop International; Lynn Cywanowicz of the National Aeronautics and Space Administration; Anna C. Urband of the U.S. Navy; Bee Falk Haydu of WASP; Col. Frank Borman of Eastern Airlines; her editor, Jeanne Gardner; Diana and Bill Gleasner, Greg Huffman, James Melton, Jim Vaughan, Anne Walkley, Eleanor Brawley, William A. Weaver and all her writing friends.

Contents

Foreword, by Frank Borman   9

Introduction   11

Tina Marie Skrzypiec, Boom Operator   21

Jill Elaine Brown, Cargo Pilot   39

Linda Elaine Barber, Aviation Inspector   53

Sandra Williams Case, Corporation Pilot   61

Mary Ellen Kraus, Air Traffic Controller   75

Joyce Carpenter Myers, Aeronautical Engineer   87

Ann Orlitzki Smethurst, Captain, U.S. Air Force   99

Margaret Rhea Seddon, Astronaut   115

Judy Ann Lee, Flight Engineer   133

Chronology   145

Glossary   153

FOREWORD

AVIATION IS CERTAINLY a field where the female presence is growing and will continue to expand in years ahead. Traditionally, women in the airline industry filled only those positions characteristic of female roles in the corporate world—clerical jobs, in-flight stewardesses, ticket agents, reservation agents, etc. In other words, women were concentrated in genteel endeavors with a minimum of physical exertion.

As we're all aware, a significant change has occurred over the past several years. Women are seeking and earning a "bigger piece of the action" with a determination not seen since their quest for voting rights sixty years ago. At Eastern Airlines, for instance, women now occupy a broad variety of roles ranging from pilot to skycap. We have female engineers, mechanics, aircraft cleaners and ramp service workers. For the first time, our marketing design office has included specially tailored uniforms for women in jobs previously handled exclusively by men.

All of this activity merely illustrates the opportunities awaiting young women today in the field of aviation. Our industry is one that touches almost all sectors of our society. It is also an increasingly complex business reliant upon advanced technological support. We have a large demand for both skilled and unskilled workers.

It is interesting to note that of the 39,000 employees of Eastern Airlines, only about 6,000 are pilots. This illustrates the tremendous support requirement in an operation like a scheduled airline. One of the most rapidly expanding elements of support is the field of computer sciences. Every function in keeping a fleet of 250 jet transports aloft each day requires computer support. The need for programmers, technicians and other computer-related skills is great and growing.

Another area where women are becoming more evident is in the operation of our terminals and sales offices. Not long ago this was a purely male domain. But now you will find women directing large teams of passenger and aircraft service personnel at major stations on our system. And you will find females carrying the major responsibility of directing sales activities for entire marketing districts.

I am pleased this situation has evolved. America was founded as a land of opportunity. Our Constitution specifies that all people will be equal in taking advantage of this opportunity. It appears that the young women of today are making a reality of this, the very cornerstone of the American dream.

FRANK BORMAN, PRESIDENT OF EASTERN AIRLINES
COLONEL IN U.S. AIR FORCE, RETIRED
FORMER ASTRONAUT AND MAN ON THE MOON

INTRODUCTION

THE FIRST WOMAN to fly solo in a powered aircraft performed her feat one sunny afternoon while on holiday in Paris. Then ''like a good girl'' she promptly allowed both herself and her accomplishment to be plunged into oblivion.

The startling news swept through Paris in less than a day and was destined for the journals across the sea. But the horrified family of the pilot, Aida de Acosta, a New York debutante, demanded a news blackout. After all, they reasoned, a real lady permitted her name in the news only twice—when she married and when she died. And what self-respecting woman would be bold enough to invade the male domain of aviation?

Aida, daring enough to sail solo above the treetops, lacked the bravado to collect her just rewards. She packed her picture hats and ruffled petticoats in a steamer trunk and returned unheralded to New York for a whirl of social activities more befitting a debutante.

Aida's flight on June 29, 1903, was in a Dumont dirigible, a lighter-than-air balloon, with a three-horsepower engine. Six months later the Wright brothers made their spectacular twelve-second flight in a different sort of craft, a heavier-than-air plane. And aviation entered a new era.

Aida's sad plight may well have hinted at what the future held for women in aviation. Although the sky stretched from horizon to horizon, it really belonged to men. The women who, like Aida, chipped off a tiny sliver of blue did so knowing they not only risked their lives but their reputations, as well. And when aviation became an industry in the late 1920s, its career positions fell like plums to men. Not until the last decade have women found the doors swinging far enough ajar to admit them to any professional status other than a few occupations of service. Still they dared.

In 1910, while only a scant handful of men were climbing into flying machines, Baroness de la Roche gained a pilot's license in Paris, France. The following year an American, Harriet Quimby, slipped off to an aviation school on Long Island, New York, for secret lessons at sunrise. To conceal her identity, she dressed in a tight-fitting aviator's suit and wore a hood that hid her face. A few weeks later a forced landing plunged Harriet onto the front pages of metropolitan newspapers. Her secret was out.

But instead of retreating meekly to her editorial job at *Leslie's Weekly,* Harriet took her public exposure as a challenge. On August 1, 1911, she became the first American woman to earn a pilot's license (there were thirty-six licensed men). The next year she soloed across a fog-shrouded English Channel with only a compass as guide, another first for women.

That same year in Paris, Mlle. Jeanne Herveu declared she'd never marry, because it would be unfair to her husband. She was young, pretty and popular. But she learned early that her own ambition would stand in the way of marriage. She flippantly embraced the propeller of her plane and announced she was already married to "this gay blade."

Perhaps Jeanne was thinking also that marriage to a flier could turn a husband into a young widower. Flying in those days was dangerous, the mortality rate high. When the early pilots referred to their vehicles as "crates," they weren't just being humorous. The planes were frail and terrifying. Wings were covered with a single layer of cheap, unbleached muslin, stretched and fastened with tacks and bamboo splits. Their edges were laced with piano wire. There were no fuselages and no cockpits. Pilots climbed into their open seats and exposed themselves to raw wind, weather, noise and whatever debris the atmosphere produced. Moving from stratum to stratum with drastic changes in temperature, they accepted sniffles and sneezes as part of their fee for flying.

The fashion of the day for women was a long, billowing skirt with ruffled blouse, not a likely costume for climbing into a flying machine. Further, the skirts often got caught in the controls and posed a danger. As time went by, women learned to shorten their skirts and to add metal guards to keep them out of the way. This arrangement was cumbersome at best. But when the women fliers switched to golf breeches, knickers or divided skirts, they were ridiculed.

Even as late as the 1930s, Amelia Earhart, who in 1932 became the toast of the aviation world by soloing the Atlantic, was criticized for her apparel. Because she wore breeches, a leather jacket and helmet, she was accused of purposely imitating Charles Lindbergh to capitalize on his fame. It's true she was tall and slender, had blond curly hair and blue eyes like Lindbergh. But these were facts of nature. Amelia cared nothing for appearance. If her clothes were like Lindbergh's, it was because they were the sensible thing to wear. But the public made Amelia so uncomfortable that Anne Morrow Lindbergh, who had made a name for herself in gliders and as copilot and navigator for her famous husband, had to come publicly to Amelia's defense.

Many of the early fliers were self-taught in a comic process they called "cutting grass." They hopped across pea patches, rose several feet and plummeted. Then they jumped out and helped repair the plane for the next trial run. Flying schools cropped up across the country, but many of them refused to teach women; or their operators sneaked away to telephone their husbands to gain permission. The women who were accepted were not permitted to work at the school to pay tuition as men were. Even after they qualified and were licensed they faced only two options. They could open a school of their own and hire themselves as instructors. Or they could occasionally get assignments for aerial photography, lug heavy camera equipment aboard, zoom to high altitudes and somehow juggle a rudder, an oxygen mask and a camera with only two hands.

Aviation remained a sport, and Americans caught the fever. Teams of barnstorming fliers zigzagged across the country putting on air shows and selling rides at any available field. They dipped and nosedived, looped-the-loop, roller-coastered and flew upside-down. As if that wasn't enough, the fliers left their cockpits and wing-walked, hung by their knees or teeth, swapped planes in midair. Phoebe Fairgrave Omlie even learned to dance the Charleston in flight and to stand upright on the top wing, supported only by piano wire, while the plane looped-the-loop.

On the ground an enthralled audience gulped and gaped and begged for more. When they swarmed around the fliers after the show begging for autographs, one thing quickly became evident. Women stunt fliers were the main drawing card. So in a weird way, by risking their necks, women could finally earn a few dollars in the air.

Then came the days of air races, derbies and daring parachute jumps. Airplanes were constantly being upgraded, and pilots were clamoring to set records in speed, altitude, endurance and distance. Records made one week were surpassed the next. Women at first staged their own races, but gradually they were permitted to race with men. In 1936 Louise McPhetridge Thaden roared to a first-place victory in the famed Bendix Trophy Transcontinental Race before some red-faced male contenders. Two years later Jacqueline Cochran won against an even larger field.

By then women were in the air for other reasons—as stewardesses. In 1930 United Airlines hired Ellen Church and seven other graduate nurses as Skygirls with the hope that their presence would stimulate business. Until then cold sandwiches and lukewarm coffee had been clumsily served by either the copilot or the mechanic on board. Now passengers would have their food, candy and cigarettes more daintily dished out by nurses, who could also hold their hands when they became airsick. These "Skyway Sallies" also mopped the cabin floor, wound the clocks, found lodgings for passengers and crew after unscheduled landings and fueled the plane. They had to be graduate nurses, age twenty-five or under, unmarried, weigh no more than 115 pounds, stand no taller than five feet, four inches and promise not to fraternize with either crew or passengers.

The experiment paid off, and other airlines followed suit. Then came the mad scramble to hire the most alluring young women possible, dress them in handsome uniforms and entice the businessman to air travel. Surely Beauty could lead the Beast to the ticket counter!

In 1950 one young applicant with all the qualifications was told to return home, gain five pounds in the right place (the interviewer fixed his gaze firmly on her chest) and return in six months. Despite these indignities, stewardesses have succeeded in turning their jobs into a profession. Now the age boundary has been expanded to sixty, they can be five feet, two inches to six feet tall, have a reasonable weight, be married, have families and fraternize with whom they please.

About the same time that the first stewardesses were hired, a few nonfliers became involved in a professional way. Laurette Schimoler became manager of the Port Bucyrus, Ohio, airport, the first woman to hold such a position; and Pearl V. Metzelthin was appointed food consultant with American Airlines. Within the decade, Anne Archibald distinguished herself as assistant vice-president of Pan American Airways; and Elsa Gardner, working with the U.S. Department of the Navy, rose to the position of first woman aeronautical engineer.

Although women aviators could not force their way into the cockpits of commercial airplanes, they became involved in other areas. Phoebe Omlie, who had danced the Charleston atop a wing, piloted Sarah Fain around the country to campaign for Franklin D. Roosevelt's first bid for the presidency. After his election she wasted no time in flying to Warm Springs, Georgia, to ask for a job. She was appointed liaison between two federal aeronautical agencies, the first woman government official in aviation.

Along with Amelia Earhart and other women pilots, she campaigned to get aviation benefits from tax on airplane gasoline and zipped across the country promoting air-marking for ease in navigating. In some areas the markings were so lacking that pilots had to dip low enough to read the town's name on the railroad station.

But national attention opened the possibility of national controls. In 1935 the Bureau of Air Commerce proposed grounding women pilots for nine days each month during their menstrual cycle. The pilots, however, now organized in a strong club, protested so loudly that the proposal was finally filed away and the matter brought to a close.

While aviation was rising to new commercial heights in the United States, Hitler was organizing his forces in Germany. The threat of another war hung heavy as storm clouds. Ironically, war seems to advance the cause of women, to open opportunities.

Prior to World War I Katherine Stinson at age nineteen flew across the country enlisting members for the Red Cross. Then she and her sister Marjorie opened a flying school in San Antonio, Texas, where they trained dozens of English and Canadian pilots rushing off to France to join the war.

During the same period, Ruth Law crisscrossed the country advertising Liberty Bond drives to finance the military. For thanks she was charged with being a spy and had to publicly defend herself. Alys McKay Bryant, an able pilot, offered her services to the government but was refused. So she went to work in a factory constructing balloons and dirigibles for the war effort. Yet none of these women had even been allowed to vote in this country, a privilege not granted women until 1920.

Following the lead of their patriotic predecessors, the pilots of the 1930s rose to the challenge. Not all of them felt the same about war. Laura Ingalls, for instance, dropped leaflets over the White House for the Keep America Out of War Committee and was grounded. Ruth Nichols, a Quaker pacifist, helped organize and became director of Relief Wings, an aerial mercy organization. But most of the others stood ready to support any war effort the country made.

In 1939 the Civil Aeronautics Administration began a program of pilot cadet training in colleges. Women were admitted, but in 1941 they were dropped to make room for more men. As men left for war, women moved into the positions they vacated at airfields, some of them even teaching cadet training. Those who couldn't find flying jobs went to work on assembly lines in airplane factories. Their pay was only 60 percent of their male co-workers' check for the same jobs, but this was wartime and women refused to grumble. Before they signed on they were told they would have to abandon their jobs as soon as a man returned to fill it. And often the factory wouldn't hire a woman unless she had a brother, husband or father already working at the plant.

In commercial aviation women were promoted to fill vacancies in reservations, sales work and even in meteorology, but these women usually pulled the graveyard shift.

Still the future continued to brighten for women. In 1941, Stephens College in Columbia, Missouri, offered a course for women in aviation, the first of its kind in the country. But the curriculum glaringly omitted flight training. The college could foresee no woman in the cockpits of the future.

The same year the Civil Aeronautics Administration was badly in need of air traffic controllers and sent out a call for applicants. However, the Administration said a flat No to women. After all, what woman could keep her head in heavy traffic?

By far the most fascinating chapter of women in aviation is one whose drama is still unfolding. In 1942 Jacqueline Cochran, along with Eleanor Roosevelt, pled with the Army Air Corps to allow qualified women pilots to serve their country. The result, after yards of red tape had been unraveled, was the Women's Auxiliary Ferry Squadron under Nancy Harkness Love, later coordinated into the Women's Airforce Service Pilots (WASP) under the direction of Jacqueline Cochran. For more than two years 1,034 WASPs flew seventy-seven types of warplanes, from single-seat fighters to four-engine B-29 bombers. They ferried the planes from production plants to stateside military bases, made test flights in new, untried aircraft, towed targets for male gunners using live ammunition. Although they were promised militarization from the start, these brave young women were abruptly disbanded in December 1944 the way they started—as Civil Service employees. The WASPs served gallantly under adverse conditions, took the same military training required of male cadets, but were paid $50 per month less than their male counterparts. To add insult to injury, the women paid their own transportation to air bases, supplied their own room and board, and bought their own uniforms, none of which was required of men. They were denied military insurance, and their private policies were cancelled because of the dangers they faced. For a final indignity, the families of the thirty-eight WASPs killed in action were not permitted to display the Gold Star denoting a war dead.

As a happy, if belated, ending, on the eve of Thanksgiving 1977 the 850 remaining WASPs, now in their fifties, sixties and seventies, were declared veterans of World War II and granted honorable discharges—thirty-three years late. Now these women can be buried in a veteran's cemetery, be given veteran's preference when seeking government jobs, and receive treatment in a veteran's hospital, none of which are retroactive. And they missed out on education benefits entirely.

During the same war, 26,000 women served in naval aviation as Women Accepted for Volunteer Emergency Service (WAVES). They were metalsmiths, mechanics, technicians, meteorologists and specialists. But none were allowed to fly.

Aviation after World War II took off like a meteor. Propeller planes gave way to jets; and the new, sophisticated aircraft became too expensive for private ownership. There were practically no jobs available for women fliers and no way to train in superjets. Jacqueline Cochran in 1953 became the first woman to break the sound barrier, but her access to the F-86 Sabrejet stemmed from her own influence and the affluence of her husband, Floyd Odlum, rather than a privilege granted a woman.

Gaining the cockpits of commercial aviation was a crawling process. In the late 1950s the Soviet Union employed its first female commercial pilot. In 1961 Turi Wideroe signed on with Scandinavian Airlines as the first woman commercial pilot in the free world. But it wasn't until 1973 that the United States followed suit when Frontier Airlines hired Emily Warner as copilot. Even today there are still fewer than 200 women pilots with major American airlines and more than 35,000 male pilots. In 1976 Emily Warner was made the first female captain in American commercial aviation and remained the only four-striper for two full years.

In the mid-1970s the military relaxed its traditions and admitted women to flying academies and flight training. And in 1978 the National Aeronautics and Space Administration (NASA) added the first female astronauts to its ranks to serve as mission specialists. As yet no American woman has been trained to pilot a spacecraft, although in 1963 Russian cosmonaut Valentina Terechkova posted a record with her forty-eight orbits of earth.

A visit to the passenger terminal of a city airport affirms today's opportunities for women in aviation. You see women behind ticket counters and reservation desks, crisply uniformed and efficiently carrying out their duties. You hear their voices on the public address system, even in foreign languages at international airports. Behind closed doors are women at desks, employed either by airlines or airport authorities as secretaries or administrative assistants, writers and photographers to staff in-flight magazines or issue press releases, lawyers and legal assistants, sales and marketing personnel, nurses or medical technicians at first aid stations.

Near the runways you find women mechanics in grease-streaked coveralls, readying the plane for takeoff. Or refuelers consulting with flight engineers. You see women rolling steps to the airplane door, or with flashlights directing a taxiing pilot. They handle luggage, serve at inspection stations and information windows, and operate parking facilities.

Aloft you find women as flight attendants, helicopter pilots between airports or, if they're lucky, one of the handful admitted as pilots to the cockpit of the transport.

Far away from the airport there are thousands of behind-the-scenes jobs in aviation, all now available to women with the capability and training: architects who plan the buildings, aircraft manufacturer technicians, designers and decorators who make the interiors of airplanes and airport facilities attractive and functional, fashion coordinators who plan uniforms for airlines.

In the realm of foods there are dieticians who work magic in arranging appealing meals to fit on a lap tray, the organizers who order supplies, the cooks who prepare the dishes.

And in control towers women are proving daily they can handle the challenge of directing air traffic.

Most of these functions are coordinated through a massive computer system, requiring thousands of technicians and programmers.

As one spokesperson for the Federal Aviation Administration put it, there are jobs for women in aviation embracing every career field except some of the arts. "We're not hiring ballet dancers or vocalists yet," she says, "but who knows, maybe someday we will."

The history of women in aviation encompasses more names, situations and contributions than can be listed. From the early balloonists to the modern astronauts spans two centuries of inspired visionaries. They ranged from hobbyists to professionals, from fliers to ground personnel. And they all had one thing in common. They refused to have their wings clipped.

The heroes today are no longer the daredevils who looped-the-loop in open cockpits or leaped midair from plane to plane. But they're heroes just the same.

Such as Nettie Dickinson, a nurse who took up flying at age fifty-three, opened a bank transport air service with her son and catapulted a one-plane operation into a half-million dollar business.

Or June Rodd, who early every morning flies out over the Atlantic Ocean to spot schools of fish for commercial and sports fishermen.

Or Peggy Kathman, an instrument flight instructor, who by telephone literally "talked down" a lone pilot trapped above clouds, untrained for instrument flying, and running out of fuel.

And the list is growing.

TINA MARIE SKRZYPIEC

☐ BOOM OPERATOR ☐

WIND WHIPPED AT the legs of Tina's olive drab coveralls as she sprinted from the Alert Facility toward a waiting KC-135. The beeper in her breast pocket was positioned to "On." The other seven pockets of her flight suit bulged with the bare essentials for bailout, survival supplies for land or sea. Knives, gloves, pen and pencil, small rations of food.

Even her underwear was geared for the ultimate emergency. Made of cotton, it was treated to resist flame, or even a spark of fire, in case of a crash landing.

Tina's senses sharpened against the bite of chill as the screeching Klaxon beat against her eardrums. In less than five minutes from the start of the signal, she must be inside the sighting window of the boom pod in the tail of the plane, ready for takeoff.

She quickened her pace.

Tina had made this short dash dozens of times—in the middle of night, in the glare of noon, at dusk when the horizon was dusted with orange. Each time the flight had been only a practice run. But the threat of nuclear attack hangs heavy in the military. This could be "The Big One."

Tina climbed up the ladder, hand over hand, to the door of the plane, grabbed a rope and pulled herself aboard. Inside the plane she secured the hatches, lifted the ladder, dropped to her belly in the glass bubble in the rear. The other three crew members rushed to their posts; the engines whirred. But not even the pilot knew yet whether the alert was a call to action or only an exercise.

Suddenly the intercom crackled. A commercial plane had been hijacked with Americans aboard. Tina's KC-135 was tasked to trail the plane, six miles in the rear, to whatever destination the hijacker chose. At the moment it was headed toward Cuba, final destination Iran. An international crisis could be ignited. Stand by!

Tina's plane was committed.

With the throb of engines against her abdomen, she lay in readiness. She rechecked her instruments, took a new reading of the information on her panels, prayed for a calm to wash over her. Gradually the dread of nuclear war fell away as Tina concentrated on news of the hijacking.

And waited.

Then the intercom crackled again. The hijacker had been subdued, the incident aborted. Tina's KC-135 was detasked.

Relieved, she returned to the Alert Facility for debriefing.

Walking behind barbed wire along a row of look-alike buildings spaced as evenly as teeth, Airman First Class (A/1C) Tina Marie Skrzypiec (pronounced Skrip-ik) looked unusually small. She isn't. Her solid 150 pounds are molded smoothly into a five-feet, eight-inch frame. Her eyes, dark and round as walnuts, looked upward each time a jet lifted in a burst of energy, drowning her voice.

"I never get enough of watching," she said. "Some of those fighters pop straight up over the buildings."

Vapor trails had already turned the sky into a scribbled blue note pad. Every two minutes another jet ripped from the ground in a thunderous takeoff; but at Seymour Johnson Air Force Base in North Carolina, such sorties are commonplace. It is one of the most important bases in the country and the only one combining both Tactical and Strategic Air Command forces. A crisis anyplace in the world is radioed to Seymour Johnson within minutes. This fills Tina with awe and pride.

"It's more than patriotism, although I'd do anything for my country," she said. "What's important is being a part of something bigger than yourself, something you really believe in."

Tina is a boom operator (air refueling specialist) with the Strategic Air Command, one of only nine women holding such positions. At age twenty she is one of the youngest.

Until recent years, the refueling of airplanes was primitive at best. To ferry planes from the United States to bases overseas during World War II was often like playing touch-tag. Pilots hopped from island to island to gas up. Delays were frequent and costly. In actual battles the bombers and fighter planes could complete only short missions; otherwise they'd face the prospect of spiraling to the ground in enemy territory, their fuel tanks dry as a bone.

Now planes can be refueled in the air without loss of time. In an intricate process, Tina can transfer (pass off) up to 6,000 pounds of fuel a minute while actually flying a mission.

Sprawled on her belly at the three-foot sighting window, she cautiously guides the boom into the receptacle of the receiving plane. When the boom nozzle latches and positive contact is achieved, she alerts the copilot to activate the fuel. Then the two planes fly in tandem until the offload is completed. The transfer can take from one minute to a quarter of an hour. Then Tina removes the boom, returns it to normal position inside the plane, signs off her radio.

View of KC-135 refueling a bomber in flight (U.S. Air Force photo)

But the job is dangerous!

"When we're on a mission, we're flying about 30,000 feet at fast speed," Tina said. "The receiving plane zooms in under us, so close I can almost touch it. We're only ten or fifteen feet apart, and it can really get sticky. We could collide, be blown or sucked together."

The pilot in most receiving planes cannot see the operation, since it takes place in the rear of his plane; so Tina takes over radio control and becomes that pilot's eyes.

"From the time the receiving craft is in precontact location to the time of disconnect, it's all mine," Tina said. "I have to stay on top of the situation, alert every minute. The possibility that the receiver would make some drastic movement is always present."

The job is easier with bombers than with fighters. The receptacles of bombers are equipped with slip-way doors, visible even at night. But with fighters the process is like threading a needle in the dark. The opening is a four-inch hole, exactly the size of Tina's boom. Fighter planes are lighter and more maneuverable than bombers. They shift and sway, making the process even more difficult and dangerous.

View from Tina's window shows boom extending from her plane to the B-52 bomber below (U.S. Air Force photo)

Tina on her stomach during actual refueling while in flight (U.S. Air Force photo)

And fighter pilots aren't known for their patience. Sometimes one of them snaps at Tina, "What's the matter, lady? Can't you handle the job?"

The lady can!

Raised with seven brothers and one sister, Tina has many times slipped over the boundary and entered a "man's world." Burning with curiosity, she infiltrated a secret meeting of altar boys in the third grade at St. Pancratius Catholic school in Chicago—and was almost expelled. "They weren't doing anything I couldn't do," Tina says now. "They were just playing baseball, but it was supposed to be so secret and special just because girls couldn't be altar boys." (Today the church allows altar girls.)

As Tina grew older, she worked side by side with her brothers on family projects. "My dad always made me do the same jobs they did," she explains. "I worked in the garage, helped build additions to the house, whatever."

In high school a semester as an exchange student with a family of four in Oregon gave Tina an opportunity to explore other life-styles.

"In a small family it was so different I couldn't believe it," she says. "They'd just decide to go to the beach for the weekend and they'd go. In ours we had to plan weeks in advance."

She watched in amazement as her adoptive family "just picked up anything they wanted" in the supermarket.

"Back home, by gosh, with ten in the family we shopped for bargains," she says. "You better believe we really shopped."

In return Tina taught the Oregon family some of the customs of her Polish background. On Good Friday she tinted eggs. On Holy Saturday she packed her Easter basket with sausage, horseradish, ham, bacon, eggs, and butter shaped like a sacrificial lamb. On Easter Sunday she shared the basket with her adoptive family and explained the symbolism. She sang Polish songs, taught them to polka.

The experience proved more than just a semester away from her home in Chicago, Illinois. For Tina it became a turning point. No longer would she be content with a sheltered life within an ethnic neighborhood. The whole world stretched before her, beckoning and exciting.

Even before graduation from high school Tina considered

joining the U.S. Air Force but decided instead to work her way through Loyola University in Chicago by continuing the waitress job she'd held since the age of fourteen. But studying, working, attending classes and keeping up with her chores at home proved too heavy a schedule.

On January 20, 1978, while President Jimmy Carter was delivering his State of the Union address to Congress, Tina filled out the forms for enlistment.

"My family was supportive, my mother delighted," Tina says. "I think she envied me the chance to do something with my life other than raise babies."

On April 24 of the same year Tina set out for Lackland Air Force Base near San Antonio, Texas, "scared to death" but ready for six weeks basic training. It proved to be the lowest point of her career.

"Our technical instructor was a tiny little girl with a big, booming voice," she says. "She was always yelling, putting us down. You're already lost and don't know what to do, and all she did was yell and scream."

Stand straight. Fold it right. Eat now. Sleep now. Get in line.

Tina felt like a preschooler. There were fifty women in her flight, living elbow to elbow, doing precisely the same thing at precisely the same moment. But Tina was accustomed to close quarters back home, a constant sharing with her brothers and sister, a give-and-take attitude. She adapted.

Tina applied for "open general," indicating she was willing to train for anything from cook to an air traffic controller. Remembering the childhood thrill of watching planes at the airport near her home, she chose air traffic control.

But there were no openings.

She took a flight physical examination with good results ("I'm strong and healthy") and was approached by a master sergeant about becoming a boom operator. In a major breakthrough, he explained, the Air Force would select one woman from her flight as a test to see if women were qualified for the position.

"He told me it was the most exciting career for an enlistee in the entire Air Force," Tina says. "And that it would be such an honor to be the first woman, and the youngest to date."

To Tina, the job of refueling an airplane sounded too much like a gas station attendant. And the opportunity to become a "token woman" held no appeal. She wasn't impressed.

But the sergeant persisted. She had scored well on tests. She had the stamina to handle the physical aspects (the boom operator must secure the hatches, lift the ladder before takeoff and is in charge of loading and securing cargo).

"Finally he said I'd be sent to survival schools, which would be like camping out, and I was hooked," Tina says. She signed up.

"Congratulations," the master sergeant beamed. "Now you can lie on your belly and pass gas."

Tina ignored the crude attempt at humor and zipped off to survival school. The first was at Homestead, Florida, for water survival. To pass, she dropped from a para-sail into channel water, churning and choppy, and was left to float two or three

Tina hugs another Polish-American member of the U.S. Air Force, Edwin A. Smuda, at Castle Air Force Base in California, 1978.

hours. She got seasick. But the thrill of para-sailing far surpassed a queasy stomach.

Then on to land survival school in Washington State, where Tina experienced her greatest surge of self-confidence.

"We started out like an ordinary camping trip," she says, "with a couple cans of rations and a few supplies."

She scooped mussels and clams from a lake, erected animal snares, made a tent from her parachute and survived for days with only a knife at hand. "We ate what was available in the woods," she says. "I even ate an ant. A black one. It wasn't too bad. I pinched off its head first."

After ten days of living off the land, Tina and her group returned to base for several hours rest before undergoing the most grueling experience of her entire life—twenty-four hours in a simulated prisoner-of-war camp. The scenario was so real, the stress and horror so terrifying, that Tina had to constantly remind herself that it was only a game of war—and temporary.

Intricate details fall under the category of "classified information" and cannot be divulged. But with slight pauses and a little prodding, Tina is able to reconstruct the portion that can be told.

In the late afternoon just before dark Tina was forced to crawl into camp, be captured by "enemy" soldiers armed with bayonets and thrown into jail. Behind bars she was thrust into a series of stress situations to test her stamina. The first, and easiest, was leaning against a wall with the weight of her body pressed upon her fingertips.

"Try it," she says. "It sounds easy. Then your hand begins to ache. The pain moves up your arm, down your back and pretty soon you're aching all over."

Then the enemy tried small insults on Tina and the only other woman in her camp section. They forced the women to serve a thin tasteless soup to the men, placing Tina and the other woman in a traditional female role to determine how easily they would become offended.

By contrast they tested the men by "torturing" the women in their presence. "We weren't actually *tortured*," Tina explains. "But those men were such good actors that it looked that way. It was terrifying. We never knew if the next blow would be real."

The captors "slapped" Tina, knocked her to the ground,

Tina with one of her biggest catches

called her a whore and insulted her with remarks Tina vowed she'd never repeat. The purpose was to force the male prisoners to weaken, to offer bargaining pleas, to sign foreign citizenship papers or an allegiance to the enemy in exchange for freedom for Tina and the other woman.

"It was really painful for those guys," Tina says. "Lots of them just couldn't take it. They signed."

Then Tina was thrown into solitary confinement in a cell too small for anything but standing or squatting. Cramped, cold and hungry, she kept her body from stiffening by stamping her feet, raising her arms, bending, stretching, twisting. Mentally she measured the cell, planned what she'd do when she got out, told herself over and over that the horror was only temporary, the game would soon be over.

But the hours dragged. And she had no way of knowing how much longer she must endure. The guards all kept their watches set at different hours to confuse and frustrate their prisoners.

In horror she remembered the stories she'd heard of real prisoner-of-war camps, of a soldier who mentally built a house, inch by inch, over a period of years until he knew every nail, every blob of mortar, every grain of wood so well he could instantly bring it to mind. Tina wondered if she, too, could stave off insanity by keeping her mind active, by creating something of beauty within her own thoughts.

The simulated camp was designed for one purpose—to convince Tina and other members of the military that they should never permit themselves to be captured.

"It worked for me," Tina says. "I vowed I'd never, never get myself into such a God-awful situation again. At a real prisoner-of-war camp there probably wouldn't be enough left of a woman to mail back home in a shoe box."

The experience brought to the surface two distinct reactions, both of which surprised Tina.

First, Tina resented deeply that the men involved were treated more harshly than the women. She filed a complaint.

Second, it crystallized her thinking on military service for women. She feels there should be a draft and that women should be included.

"But not for combat," she says. "In another culture—in

China or Israel perhaps—it would be all right. But in our culture men have been taught to protect the female. In battle I think they'd jeopardize their own safety, and the safety of others, just to protect a woman.''

Tina fared well in survival, was zipped off to boom operator school in California and upon graduation was assigned to Seymour Johnson. It hasn't all been perfect.

For one thing, there's loneliness. Since she remains the only female boom operator on base, she has no other woman to share her experiences. And because she is assigned to ''crew rest,'' which means she must get eight hours of uninterrupted rest before each flight, she cannot even have a roommate.

Tina lives in a dormitory on her assigned base, in a cubicle with only the bare essentials—bed, table, locker—all government

Tina, front row, third from left, with the first group of female boom operators in history (U.S. Air Force photo)

issue, functional and drab. She has no cooking facilities other than a hot plate, no place to entertain privately.

As boom operator, she deals mostly with officers rather than enlisted personnel such as herself. She learns to know them well and enjoys their company. Yet fraternizing socially with officers is discouraged, and she must find her close friends and dates among the ranks of enlisted personnel.

Every third week Tina goes "on alert," the one assignment every crew member in the Strategic Air Command dreads. She and nineteen other crew members, all men, set up residence in a squat, rectangular building known as the Alert Facility. The facility is rimmed by barbed wire and patrolling military police. For seven days and nights Tina is permitted to move no farther than five minutes from the KC-135 plane sitting ready within the barbed wire.

The hours drag by. Tina uses the time to do needlepoint, sewing and embroidery. She and the men play endless rounds of Ping-Pong, meaningless card games, watch television, read.

And listen—always listen—for the screeching Klaxon calling them to duty in the sky.

"About the fifth day we all get kind of depressed," Tina says. "Our morale drops, and we begin counting the hours."

At first the Air Force felt women should be exempt from alert, that it was too trying, too demanding. But Tina is glad no action was taken. "If a woman can't take alert she shouldn't be in a field that requires it," she says.

Despite her positive attitude, Tina feels a constant pressure to "prove herself." "On a check ride, for instance, when I'm being evaluated, I feel I must do exceptionally well just to keep from letting down other women," she confides.

At times Tina encounters a reluctance from other squadron members. If the cargo is especially heavy the men will always help one another. But if the cargo is hers, they say "It's *your* job."

Tina's schedule is divided into periods of three weeks each. For two weeks she is on regular assignment, either flying or preparing for flight each day. On the days of preparation she joins her crew members at squadron headquarters to make charts, maps and to configure the plane, making certain all fuel, passengers, crew and cargo are accounted for.

Tina is in charge of weight and balance. Using a slip stick, she draws up a form showing placement of fuel, oil, water and every person on the plane. She orders lunches (not because she's the only woman but because it's the boom operator's job), then helps the navigator with his charts.

On flight day she arrives at headquarters 2 $1/2$ hours before takeoff. En route she picks up the lunches, coffee and water for the crew. At the plane she loads the cargo, then does a thorough preflight inspection of the cargo compartment and all the equipment associated with her job.

A refueling mission averages more than five hours and takes Tina thousands of miles from her base. On the return flight Tina assists the navigator with the celestial navigational leg. Using a sextant, she locates the sun, moon or a star. Working from this fixed point, she must find all stars in the quadrant and guide the crew home without instruments, just as sea captains did in days long past.

In crisis or in war, celestial navigation is a must so the aircraft can maintain radio silence. Otherwise the signals could be picked up by the enemy, their location detected.

When Tina first arrived on base, she was surprised to find herself in the spotlight. Sometimes the bright beams got in the way of her work.

"The commander encouraged me to give interviews for publicity," she says. To Tina he intimated it was the least she could do to repay the Air Force for the privilege of becoming a boom operator. Flights were set up for Tina and the only other woman flier on base, a copilot Tina would not normally fly with.

"That was for the press," Tina says. "I understood all the reasons, but that didn't make me appreciate being exploited."

Tina has also been asked to help recruit, to impress young women with the excitement and advantages of her job. This bothers her.

"There are about 900 boom operators in the entire Air Force right now," she says. "Only nine are women. The number will increase, but I doubt if the Air Force will have more than 25 percent when the program gets into full swing."

There are decided advantages in the Air Force, Tina says. Such as earning a college degree while in training, or drawing

Tina adjusts the tail support for the boom prior to takeoff. (U.S. Air Force photo)

temporary duty assignments to fascinating foreign lands. But becoming a boom operator requires persistence, hard work and patience to await an opportunity. Only a few will make it.

Tina, meanwhile, has proven that women can excel as boom operators. She has advanced from airman first class to sergeant, from trainee to instructor. It no longer disturbs her when a new pilot picks up her soft voice on the interplane radio and gasps, "My God, a woman!"

She now is stationed on Okinawa, a small island near Japan. Her refueling missions take her far across oceans to lands that were until now only colored spots on the globe—Korea, Guam, Australia, the Philippine Islands. Refueling over the sea fills Tina with a new surge of dedication and an overwhelming sense of self-worth.

Tina looking through window of boom pod during refueling (U.S. Air Force photo)

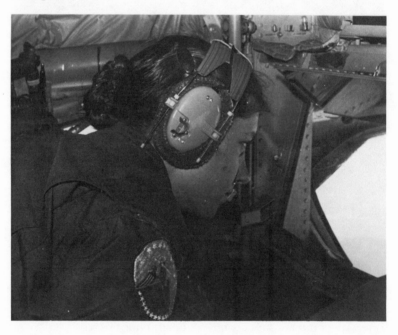

"I look down at that plane under me and realize I'm their one link to life," she says. "If I can't make contact, if I can't refuel, the outcome is obvious.

"That plane drops into the sea."

Jill climbs aboard plane. (photo by David Mark, courtesy Zantop International Airlines, Inc.)

JILL ELAINE BROWN

□ CARGO PILOT □

JILL BROWN IS a "day" person. She loves mornings, feels at peak capacity while the sun beams down its energy.

Yet several times a week Jill takes off in the darkest dollop of night, winging her way through stars and dropping off cargo at night-drenched terminals.

It's the price she pays to stay in the air, a commitment she made to herself while still a teenager. Now at the age of thirty she's one of three female cargo pilots for Zantop International Airlines—the first and only black one.

Jill's position at Zantop is not her only "first." Her most notable was the day she became the first black woman to fly for a major passenger airline in the United States, an honor that turned sour midair.

But that's another story.

In fact, Jill's life is a series of separate stories, strung together like planes in formation. At a glance they all seem to have happy endings, to make Jill look like a fairy princess with a magic wand. Actually the very things that brought enchantment to her life are the conditions that have caused her the greatest problems.

Jill is pretty, with a soft voice and a manner to match. She's immediately likable. She's confident, accustomed to being listened to and respected. She's an only child of supportive, hard-working parents. In a crowd she stands out. She wears an air of success.

But Jill's career path has been strewn with cockleburs.

She's been handled as though she were a commodity, suf-

39

fered a bad scene with the United States Navy, and now she flies by the light of the moon with cargo when she much prefers a day-light course with passengers.

Why does she stick with it? Because she, like other women in aviation, is addicted to flying. "And with the Lord's help I'll make a success of it," she says.

Jill began flying in a Piper Cub at age seventeen. When she turned eighteen, her parents bought a Cherokee 180D for week-ending and vacationing. They dubbed it the "Little Golden Hawk."

"We called ourselves Brown's United Airlines," Jill says. "I used to ask if I could use the plane like other kids asked for the family car." She flew her friends from her home near Baltimore, Maryland, to the family farm in West Virginia. Or to a fast-food chain in a neighboring state because they offered the biggest hamburgers around.

On vacations the Browns island-hopped in the Caribbean, flew to Mexico City and Acapulco, Mexico, to Puerto Rico and even as far as Trinidad.

But Jill's summers were not all dappled beaches and candle-light dinners on a tropical island. Because she had no brothers, her father taught her chores the way he would have taught a son. When Jill was nine, she operated a tractor, worked weekends on the farm. At fifteen she began earning her own money.

"I peddled farm vegetables door to door all summer long that year in Baltimore," she says. "That took a lot of door knocking."

At sixteen she became a paint contractor for her father's construction company, climbing ladders, painting houses inside and out. At seventeen she became an assistant playground director; at eighteen she worked at the Post Exchange at Fort Meade, Maryland; at nineteen she was a clerk-typist; at twenty a country club bartender; and at twenty-one she began house painting again.

"While my father was teaching me that I could do traditional men's work, my mother was stressing femininity," Jill says. On her mother's advice Jill majored in home economics at the University of Maryland. "My mother is a teacher and thinks it's the ideal profession for a woman," Jill says. "I was always kind of handy in the kitchen, so I chose home economics."

Upon graduation she took a teaching job in Massachusetts.

Jill, age seventeen, with the first plane she flew

She found it less than inspiring. Most of her students chose her classes as a "crib" course to avoid the more difficult math and science electives. They really weren't interested in home economics.

Jill continued flying with every spare penny she could manage to save. She earned her instrument, commercial and instructor's ratings. But she was frustrated. Jill had dual ambitions—one to be a professional singer, the other a commercial pilot—and neither was being realized. The singing career had been nipped some years back by her parents, who objected to the rigors of the night club circuit for their daughter. And to become a commercial pilot required more hours and schooling than she could afford.

Feeling the military would supply the answer to pilot training, Jill had checked with the Air Division at Fort Meade when she was nineteen, asking when they expected to accept women. The recruiter laughed.

Later from her teaching position in Massachusetts, she maintained close contact with the U.S. Navy recruiting center. Finally in 1974 she received a telephone call. Now that women were being accepted into the military, Jill would be a prime candidate for officer's training in the Navy Air Force. She signed up.

For Jill it was a bummer.

"In the military you must conform," she says, "just be one of the crowd." Because Jill was black and a woman, she found it hard to be "small and inconspicuous."

"My every move was watched," she explains. "And I made some bad mistakes, some really bad ones." The worst: "Not being able to keep my mouth shut."

"In the military you just keep quiet and take orders," she says. Because Jill is outspoken, accustomed to saying how she feels and being listened to, she chalked up enough demerits to warrant a conference.

"After six months, the Navy and I decided I'd just take my honorable discharge and leave," Jill says. She manages a tight smile when she speaks of it now, but it was a devastating experience.

She felt the Navy offered the best flight training in the world, and she had blown her opportunity. Even worse, the press had covered her enlistment, and she had received fan mail from blacks across the country. Now she had let them down. Cockleburs!

Dejected, she returned home.

"It was humiliating," she says. "I honestly couldn't face people. It was a long time before I'd even go out of the house."

When she did emerge, she made six trips to the airport before she mustered up enough courage to go inside to sign up for advanced training.

But Jill was learning many things, both about herself and about life in general. "I always wanted to be someone special, to do something significant," she says. "Someone once said that to be a success you must find a need, then fill that need. I felt women would someday have a chance in aviation, and I was determined to be ready for it."

She returned to the teaching profession and for two years instructed inner-city Baltimore children.

"I taught all day, then took flight training at night," Jill says. "I was just exchanging my paycheck for lessons, but I was at least progressing." She earned the advance ratings, plus a multi-engine rating.

Then one fairy-tale day in 1976 Jill noticed an article in *Ebony* magazine about Warren H. Wheeler, a black man who owned a small commuter airline in Raleigh, North Carolina. The article stated he was hauling, among other things, rats for laboratory experiments.

"My heart did a flip," Jill recalls. "I'd fly rats, anything, just to get back in the air." With a black owner perhaps she'd stand a chance.

She telephoned Wheeler and flew down for an interview.

"You have the job," Wheeler told her after a check-out flight. "I'll call you as soon as I'm able to add another pilot."

"I'll fly for nothing," Jill told him. "Find me a ground job just to make enough to live on, and I'll fly right now—with or without pay."

Wheeler, it turned out, needed a ticket seller, but the job paid only $300 a month.

"I'll take it," Jill said.

She rented a room in a private home and kept a close check on expenditures. The sacrifice was nothing compared to the joy of being in the air.

She copiloted, hauling cargo—rats, even—and flying every chance she got. Eventually she was made resident copilot in a Beechcraft 99 twin-engine aircraft, considerably more demanding than any plane she'd flown to date.

"I was up every morning and at the airport before 6 A.M. for preflight," she says. "I had to check the oil and hook up battery carts. Then I took reservations and wrote up the tickets. After that I loaded the passengers' baggage—by myself."

Then she supervised passenger boarding. After the passengers were seated Jill made the announcements and inspected the fifteen-seat cabin to be sure each seat belt was secured. Then she slipped into the copilot's seat in the cockpit.

"The time in the air was great," Jill says. "But the groundwork was miserable. The loading and unloading were hard. I had no help at all. I had to chase after run-away baggage carts in all kinds of weather, lift, pull, stretch, lug. I was worn out and felt haggard all the time.''

Nonetheless, she remained at Wheeler until she had logged some 800 hours in the Beechcraft for a total of 1,200 hours flying time. At last, enough hours to qualify for a major passenger airline!

Jill zipped out applications to the familiar lines and waited hopefully for their replies. Strangely, the line that contacted her was a new, small company, one to whom she had not yet formally applied. The officials were interested in Jill, wanted her in their training program. Unaware that she would be used as a ''token'' to gain publicity and prestige, Jill signed on in March 1978 as their first black woman in pilot training.

Six months later she washed out.

This time the cockleburs dug deep into Jill's flesh. Remembering the bitter lessons learned in the Navy, Jill had set out to become a model student. She progressed with the work and was doing well. Then one day the airline's vice-president engaged her in conversation.

"Don't you *ever* think of leaving our company," he said, a smile creasing his face. "You're a highly marketable commodity."

A *commodity!*

Stunned, Jill felt trapped and confused by her own uniqueness. First, the Navy wanted her to be ''small and inconspicuous.'' Now a commercial airline wanted to thrust her on the stage, front and center, to help launch a new company.

But all Jill wanted was an opportunity to become a professional pilot, and a good one at that. Her sex and color were certainly not commodities to be bargained for.

Ordinarily, when a training class graduates from ground school or receives Wings, it's a routine process with no fanfare. But this time special ceremonies were planned. Company officials refused to accept Jill's protest against publicity but did finally agree not to single her out from the other women. They failed to keep their promise.

The news media, notified well in advance, sent crews of photographers and scribbling reporters. Headlines and photo captions called Jill "the first black woman to qualify to fly for a

Jill during preflight checks (photo by David Mark, courtesy Zantop International Airlines, Inc.)

major U.S. airline.'' She was made to pose in the cockpit of a
D.C. 9 as if she were flying it. But in reality all she had accomplished to date was completion of ground school in an entirely
different plane, the Convair 600.

"I did fly the line in the Convair, but my instructor, a male
chauvinist, purposely withheld information so I couldn't make
it," Jill says. "I requested another instructor but was refused."
Instead, Jill, feeling used and betrayed, was quietly assigned to
another session of training. This time without fanfare.

Her relationship with the airline soured. As a "commodity,"
she had succeeded. As a pilot, she was nowhere.

Jill's self-esteem plummeted. She searched within herself for
answers. Her desire to fly still burned hotly, her ability had been
proven. Yet each time she came close to a professional success she
found her own individuality blocked her path. There was something missing in her life, some necessary ingredient to success.
And she set about to find it.

First, she knew she must prove herself again in the air, erase
the bad experiences. And second, she felt the need for some
deeper purpose to her life.

One month later she found them both.

At Willow Run Airport near Detroit she signed on with Zantop International Airlines, the world's largest commercial all-cargo carrier. At Zantop she'd copilot the Convair 640, basically
the same plane she had trained in at the passenger airline. The
technique of flying cargo is no different from that of flying
passengers. The hours are more confining, the pay considerably
less, but it was flying. And an opportunity at the controls.

Now for a place to live.

"I don't like living alone," Jill explains. "And I suppose I
was really looking for a home life." So her mother joined her in
Detroit and they set out for the suburbs.

By happenstance, she thought (although she now attributes it
to Providence), they were driving down an unpaved country road
when they spied a woman washing her car on her front lawn.

"I had never done anything like this in my life," Jill says,
"but we just parked the car, walked up to the woman and asked
if she'd like to rent me a room. She was as surprised as we were."

The woman, it turned out, had three bedrooms. Except for

Jill reports to work. (photo by David Mark, courtesy Zantop International Airlines, Inc.)

two dogs and a fish bowl of flitting guppies, she lived alone. Yes, she thought she'd like a roomer.

One evening shortly after moving in, Jill passed her land-lady's bedroom and saw her reading a Bible.

"I was so impressed I just sat and talked with her," Jill says. "I asked a lot of questions about Bible teachings, things I'd puzzled over since childhood." Over the next months their conversations continued intermittently, on Jill's way out of the house, when she came in from a flight, whenever the time seemed right.

A year later another friend invited Jill to church. It proved a turning point. The church was fundamentalist, its members friendly, sympathetic, and, according to Jill, "full of love." Jill joined.

Now for the first time her life is prayerful. On the drive to the airport, before takeoff, before landing and often midair, Jill prays for wisdom and safety, for guidance for her captain, even for the other planes zipping by. A Standard Revised Bible is routinely part of her luggage, always within reach.

"I haven't lost my ambition or changed my goal," Jill says. "The emphasis is just different." Where once she hoped to become the first female captain in aviation history, now she wants only to be a good Christian. "Anything I'm able to attain after that will be a reflection on my Christianity rather than on Jill Brown."

Cargo flying has a built-in bag of pluses all its own. Even though Jill is a "day" person, she thrills to the enchantment of night flying, the time when most cargo is moved. From her cock-pit window she looks down on a spread of lights below, twinkling like a string of Christmas bulbs—Louisville, Cleveland, Chicago. She recognizes the city by sight, each with its own peculiar pattern.

And there's nothing to surpass the joy of flying into a sunrise, touching down when the runway takes on the first pink glow of dawn.

For layovers or to conform to the legal rest period between flights, Jill stays in a motel near the airport. In summer she can loll beside the pool. In winter she reads, watches television,

Jill preparing preflight reports (photo by David Mark, courtesy Zantop
International Airlines, Inc.)

makes new friends. But always she must rest before taking to the air again.

"And I grow fat," Jill laughs. Even though she's trim as a celery stalk, Jill keeps an eye on calories. "Eating becomes such a big thing when you're away from home," she explains. "The crew begins picking out a restaurant even before we land."

She has her favorites. In Boston it's a Polynesian buffet and cheese cake at the Sheraton. In Atlanta it's pork barbecue. In Baltimore it's Maryland crab cakes.

But cargo piloting is confining. Except for one free weekend a month, Jill is on call twenty-four hours a day. Flight assignments are posted on a large board inside Zantop's Flight Following Room. When she has completed a trip, Jill's name is removed from the top and added to the bottom like a chain letter. She can predict her next flight with some accuracy, but the situation can change as quickly as Detroit's weather. Her required notice is two hours before flight, but Jill can at any time receive an ASAP (the pilots call them A-saps), which means "as soon as possible," and certainly less than two hours.

As copilot, Jill does much more than share time at the controls with the captain. She takes weather briefings, makes weight and balance sheets, gets releases, files flight plans. At each freight terminal she tackles another stack of paper work on shipments dropped off and shipments picked up.

Cargo pilots always carry a suitcase with several changes of clothes. Their business is erratic, their schedules frequently altered.

"One of our pilots took off for a three-day trip to Alaska and stayed three months," Jill says. "We're always prepared for a switch in plans."

Because cargo carrying is dependent upon the health of the nation's economy, pilots are often "laid off" during slow periods. Zantop is primarily a carrier for the automotive industry, so layoffs can be frequent. Except for unemployment compensation, the period is without pay. Jill, as one of the newer pilots at Zantop, is one of the first to be notified.

But Jill is steadily building seniority and time in the sky. At 2,000 hours she'll be eligible to train for captaincy, a position that pays considerably more and doesn't require the mounds of tedious paper work.

Sometimes it hurts, but Jill pushes into her subconscious the joys of daylight flying. At 2 A.M. on a day when she planned a swim or a hike, her phone rings and Jill crawls out of bed, packs her bag and heads for Willow Run Airport.

After all, it's flying. And that's where her heart is.

Jill preparing paperwork prior to takeoff (photo by David Mark, courtesy Zantop International Airlines, Inc.)

Linda before takeoff (photo by Greg Hubler)

LINDA ELAINE BARBER

☐ AVIATION INSPECTOR ☐

LINDA BARBER SLAMMED down the receiver, jumped into boots and coveralls, tucked her hair under a cap and raced to the parking lot.

If she were to design a likely day for an airplane crash, this would be the one. Clouds hung thick and dark as crude oil. An unrelenting wind whipped a wash of rain against her windshield.

Threading her way through traffic toward North Atlanta, Linda fought against her persistent, plaguing fear. *Would it be someone she knew?*

Linda had learned to deal with injuries when investigating a crash. With fatalities even. But they had all involved strangers, code numbers on the tail of a plane.

It was different now. Most of her friends were in aviation. She knew their planes on sight, recognized their numbers. The dread of finding one at the scene of a crash followed her like a tail wind.

Ahead she spotted the ambulance, police cars, rescuers scurrying about in orange slickers. Good. They'd already be at work, removing bodies from the tree. It helped, getting that done before Linda began her own work.

She edged her car onto a grassy strip, grabbed her inspector's kit, stepped into a slush of mud.

"Hold it," a police sergeant snapped. "No spectators."

Linda whipped out her identification card.

"Federal Aviation Administration inspector," he read aloud. "Linda Ba...a girl!" He returned the card, smiled, stepped back. "Sorry, ma'am."

Linda felt his admiring eyes on her back as she hurried to the wreckage, but she's grown accustomed to being watched. Her fragile beauty makes her look more like a candidate for Miss America than an inspector in general aviation. It's part of the reason she wears coveralls, and hides her long blonde hair.

Linda saw before her the worst wreckage of her career. Airplane fragments were scattered as far as her vision could scan. Her eyes darted from piece to piece, searching for the tail with the identifying numbers. Finally she spied it across the road, protruding from a bed of mud. Rain had rinsed away the slime; the numbers stood out bold and clear. *Unfamiliar.*

Relieved, Linda set to work.

Linda has learned to position her emotions on "hold" while doing the actual investigation. At first she choked back nausea, struggled to steady her trembling fingers. And inevitably the scene returned to haunt her dreams night after night. But with effort and experience she finds she can be objective during the process, view hers as a job that must be done. And there's always the possibility that her findings could prevent a future tragedy.

"You really have to develop a strong stomach," she says. "I've actually stepped on disjointed fingers. And found part of a human head on the instrument panel."

It's Linda's job to rope off the area, locate every piece of wreckage, climb into the cockpit to verify the radio frequency, read the instruments, check the fuel and note every inch of damaged material.

Then she must examine the location, comb the terrain, scrutinize the trees and bushes to detect how the plane came down, what it struck. Her investigation can run into days.

Eventually Linda will file a formal report suggesting the probable cause of accident, a responsibility that would hang heavy on an average twenty-five year old. But Linda loves the challenge.

In this case the pilot broke suddenly through thunderclouds, realized he was nosing toward the ground and pulled back on the controls to correct his position. At his excessive speed the plane was overstressed. It ripped apart.

Linda is one of the youngest inspectors with the Federal Aviation Administration and one of the few women to hold such a job. Until three years ago she was a secretary in the same office near Atlanta, Georgia, taking messages and typing reports for male inspectors. Her meteoric rise in one giant step may sound like a Cinderella story. Actually Linda, like the heroine of the fairy tale, missed a lot of fancy dress balls. She moonlighted at night, studied during coffee breaks and lunch hours, worked at nitty-gritty jobs and sold many cherished possessions (including her motorcycle) to get where she is.

Linda's intrigue with aviation began in high school, when a boyfriend took her flying. But it wasn't until she went to work as secretary at the Federal Aviation Administration in 1974 that she was really smitten.

"I talked with a lot of pilots, and it all sounded like so much fun," she says. "They all seemed so in love with their jobs."

One day a few months after she began her secretarial job a flight instructor offered her a ride in the sky after work. She went up—and fell in love with both flying and the instructor. She signed up to take flight training with her new friend. Two months later she married him.

Linda at the controls (photo by Greg Hubler)

But the marriage was short-lived. Her husband, an excellent pilot, was a perfectionist. "He expected me to give a perfect flying performance every time we were in the air," she says. "Sometimes I'd just end up crying."

She gave up marriage but continued her lessons with someone else. Linda learned early that flying, because it is so costly, involves a host of sacrifices. "I sold anything I could get my hands on," she says. "Then I took out a big bank loan—and I'm still paying on that." She earned her instructor's license.

Then she began instructing after work to pay for more lessons and to whittle off some of her steadily mounting debts. Flying exacted a steep price. But Linda was willing to pay.

"I'd get up at 5 A.M. and teach until it was time to go to the office," she says. "Then I'd teach until 9 P.M. Sometimes I'd still have another job, flying copilot for a company or ferrying a plane somewhere."

There were nights when she slept in her car near the airport, then walked across the parking lot to get to her office. On weekends she instructed from dawn till dark. In less than two years Linda earned a commercial pilot's certificate with single engine, multi-engine and instrument ratings, as well as a flight instructor's certificate. And she was constantly building up hours in the sky to qualify her for other certification.

But Linda's plans took a nose dive in 1977 when the Federal Aviation Administration decided her moonlight instructing constituted a conflict of interest. Grounded and confined to a desk job with no funds for flying, Linda felt victimized. She considered resigning, but before she took action she received a notice to appear at the personnel office.

"We've been going over your records, and we like what we see," the director said. "How'd you like to become an inspector?"

Linda skyrocketed! This was exactly what she wanted but never dreamed of gaining. The promotion from a desk job to an inspector was unprecedented. And she was so young—and a woman!

"I think now that's what they had in mind all along when they restricted my moonlighting," Linda says. "I just didn't realize it."

As inspector, she would not only have access to an airplane to keep her credentials up to date, but she would be testing pilots for licensing. Time in the air!

Linda zoomed off to the Federal Aviation Administration Academy in Oklahoma for three months of concentrated training and returned as an official inspector. Her job has several facets, all dealing with general aviation as it applies to the state of Georgia. She investigates accidents; gives flight tests to pilots when they apply for certificates or ratings; checks out incidents (minor accidents), violations, complaints; inspects pilot schools to determine if they are operating up to code; and verifies air taxi operators.

Each responsibility involves potential danger. When Linda checks out a student, the student is at the controls. Linda watches him carefully as he goes through standard maneuvers outlined by the Administration—a spin, a stall, a turn—whatever Linda assigns. The pilot may become nervous and do a bad job, and Linda must take over the controls.

"Once we take over that's an automatic failure," she says. "There's always the chance some crackpot might lose his cool and

Linda (right) with student (photo by Greg Hubler)

threaten me. He's bigger, so all he'd have to say is 'Okay, lady, you fail me and I'll kill both of us.' It sounds like movie stuff, but it could happen.''

When Linda fails a pilot, she tries to do it as painlessly as possible. She knows only too well the hard work, study and money it takes, and she's sympathetic. But she's also conscientious and has no intention of approving a pilot not ready for the controls. Sometimes the male students flirt and offer a dinner date in exchange for another chance.

Linda has no patience with pilots who take unnecessary risks. She recalls a recent incident of her own. While moonlighting she was scheduled to copilot a past-midnight cargo flight. But her day had been long and particularly difficult. She didn't feel at peak, so she cancelled. The pilot flew out at 1 A.M., crashed on takeoff and was burned to death.

"That completely tore me up," Linda says with a shudder. "It still sticks in my mind like glue."

The plane was overloaded, which Linda finds inexcusable. Every pilot should know the capacity of the plane, she says.

Other dangers could stem from reports Linda must file on incidents, complaints or violations. She passes these reports along to the legal department of the Federal Aviation Administration, along with her recommendations for whatever suspensions, fines or censure she feels warranted.

Recently, for instance, a pilot was flying instrument in bad weather, had a radio problem and lost touch with the air traffic control tower. He suddenly broke through the clouds, saw a runway and proceeded to land. Without communication, the tower reported the plane lost and sent out a search team.

"I investigated this incident and suspended that pilot's license," Linda says. "He should have telephoned the tower as soon as he landed."

The pilot threatened to sue.

"If I were the one to check him out for his next test after his certificate is reinstated, he may very well try tricks to scare or threaten me," Linda says.

To protect their inspectors, the Federal Aviation Administration is careful to avoid giving out home telephone numbers and addresses.

Probably the most realistic danger is on accident reports. She must enter every downed plane, regardless of its position or condition. "That craft is often tilted and could easily flip over with me inside," Linda explains.

The purpose of an inspector is to promote safety, not only for those in the air but for those below. But inspectors are often looked upon as the "bad guys." Sometimes when Linda arrives to investigate a school operation or a complaint, she'll hear someone announce, "Hey, here come the Feds."

But Linda's biggest problem with people is getting them to realize she has the same authority a be-spectacled, sixty-year-old male inspector has. "I used to be shy about it and not say

Linda preparing to examine crashed plane (photo by Rupert L. Powell)

much," Linda says. "Now I've learned to speak up, be firm, make people obey."

Linda works at playing down her femininity on the job. She wears long pants or tailored suits, is careful not to be overly friendly. When she goes on inspection trips with male inspectors, she carries her own bag, opens her own doors and gets her own meals. "Fortunately the other inspectors treat me like a pro rather than a woman just tagging along," she says.

Sometimes when fliers come in for a test and see Linda in charge, they complain. "I don't want a secretary," they'll say. "I want a real inspector." One pilot vowed he'd never flown with a woman and never intended to. He rescheduled his flight check.

Linda's confidence grows daily, and daily she learns more about herself and her job. Looking back, she is often surprised at her own display of determination.

Linda is buying a two-story brick home in the suburbs. She does her own cooking and cleaning, her own yard work. In creative moments she sits down with her macrame or writes a poem. When she's feeling outdoorsy, she heads for the slopes for some skiing in the winter, or to the shore for scuba diving in the summer. Her latest interest is karate, which she's learning for self-protection.

Meanwhile, Linda continues to build her flying time for greater opportunities. She's at 1,500 hours now. At 2,000 she can expect jobs that will give her more time in the sky.

"You have to be a pilot to understand, but once you start flying you get addicted," Linda says. "There's no way in the world I could ever see Linda Barber not flying. If I go even a week and a half without getting into some kind of airplane I get real itchy."

The job pays well, but Linda never thinks of it in terms of money. To her it's a means of staying in the air.

"I was at a training seminar the other day," Linda says, "and was told to write down the five most important things in my life."

Writing it was easy, she says. Airplanes. Career. Family. Friends. Health.

Then she was told to give one thing up.

"That's when I found out what aviation really means to me," Linda says. "I couldn't possibly give up flying."

SANDRA WILLIAMS CASE

☐ CORPORATION PILOT ☐

"ONE DARK AND stormy night," her father began, his voice low and quivery, "I was driving down a desolate road in a hearse as black as midnight."

Sandy snuggled into the corner of the sofa, her knees tucked under her skirt. Her father was the greatest storyteller she knew, and this was her favorite tale, just frightening enough to make a preschooler feel brave for listening.

"There was nothing to hear but the splash of rain, the moan of the motor," he continued. "For I was all alone in the darkness —except for the dead man behind me."

Each time Sandy asked to hear the story her father would add another dimension to make it more scary than ever. Hooting owls, swaying shadows, wind whistling around the windows of the hearse. When he came to the part where the automobile got stuck in the mud and her father was trapped for the night with a corpse, the goose bumps on Sandy's arms felt as big as jelly beans.

"*Then suddenly the dead man moved!*"

The night of a telling Sandy would crawl into bed with her mother and father. And she'd wait weeks before asking to hear the story again.

Years later, when a grown-up Sandra Williams Case was trying to land a job as a charter pilot, she vowed she'd fly any cargo in the world except one—a corpse. And she's done just that. She's flown caged animals, tropical fish, dogs and cats. On one assignment she flew detonators into the teeth of a storm. The de-

Sandra with pilot before takeoff

tonators were unpackaged in the cargo pit behind her, rolling and bumping against one another. With every quiver of the plane Sandy feared her cargo would explode and she'd be sent streaking across the sky like a jag of lightning.

But when it came to transporting corpses, Sandy stuck to her vow. The other pilots teased her, but one of them always obligingly took the flight.

Sandy can laugh about her hang-up now. The days of cargo flying are far behind, and her friends have stopped mentioning it. Instead they look at her with a considerable amount of awe. And for good reason. In 1977 Sandy signed on as a pilot for General Motors Corporation, an enviable position for any flier. But even more impressive is the fact that she's the first woman pilot to be hired by this giant corporation. Now she zigzags across the country in shiny jets, transporting corporate officials to and from conferences of international importance. One day she may lunch in Rochester, New York. That same evening she may sleep in Beckley, West Virginia. She flies to Bermuda and Mexico, to Labrador and Canada, where once it was so cold she had to re-

Sandra with pilot before takeoff

move the battery from her plane and check it into her hotel room for the night to be assured of starting the next morning. Her next trip was to Palm Springs, California, temperature 110 degrees in the shade.

Sandy is affable and good natured, so her friends still tease her. But now the teasing is about another aspect of her life—her trips to the hairdresser. Every three months Sandy hops a plane from Detroit and flies to a tiny town in North Carolina to get her hair frosted, a round trip of more than 1,000 miles. This extravagance may make her sound like a jet setter. It's true she has the looks, the style, the *savoir faire* to be just that. But she isn't.

As a career pilot for one of the world's largest corporations and the single parent of two teenagers, Sandy has no time for style setting or a fast social pace. Instead she keeps her home in order, sitters lined up for her children, her dark blue flight suit cleaned and pressed, and her bag packed. Even in her leisure hours she remains constantly in touch with her telephone for a sudden change in schedules or an emergency flight while on company trips.

The momentous transition from her first ride in a small private aircraft to the pilot seat of a 500-mile-per-hour jet swallowed a twenty-year chunk of Sandy's life. Today a woman can conceivably make the same journey in three to five years.

When she was eighteen, Sandy left her home in Akron, Ohio, to enter Ohio University in Athens. Her mind whirled like a propeller as she examined the curriculum. Sandy still had made no career decisions, so choosing her electives was difficult.

Then she spied the flying program.

"I knew instantly, right on the spot, that I wanted to make a career of flying," she says. "Something clicked."

Sandy didn't learn until later that her father had zoomed around the country in open-cockpit planes during the early days of aviation. He, too, may well have chosen flying as a career; but commercial aviation was a dream of the future in those days. Now, five decades later, it seemed a natural for his daughter.

Sandy enrolled. But before her freshman year was over her father became terminally ill. She returned home to share his last few months of physical life and transferred as a commuter to nearby Kent State University. In the campus bookstore she plunked down money for the books she'd need for their flying program and set out to sign up.

But she was barred from class.

"It was a shock," Sandy recalls. "I really couldn't believe it. Kent was a large, progressive university, and this was the early 1960s. They'd let a woman take flying lessons but wouldn't let her enroll in the aeronautics program for a degree."

So Sandy reluctantly grounded all thoughts of cockpits and sky travel for the time being and concentrated on getting a degree in her second choice, Spanish and psychology.

In her junior year she married and shortly thereafter began her family.

"It took me four years to complete my senior year," Sandy says. During these years, she maintained a home, had two children, resumed her flying lessons and went to night school. "It was hectic; but by the time I graduated, I had earned my private pilot's license, plus a degree in Spanish and psychology. And I knew I wanted to stick with flying."

It wasn't a happy decision for everyone. Her husband, Tom,

was employed in a family business, had ample income and wanted her to stay home with the children. Her own mother feared for her safety, and Tom's mother was so incensed she refused to speak to Sandy for months.

But even in enemy camp there is often an ally. Sandy soon found her ally was her father-in-law. He thought flying was a neat idea, Sandy says. He encouraged her to stick with it.

And although Tom objected to Sandy's flying, he provided funds for lessons. Sandy reimbursed her husband by teaching both him and his father to fly. Tom's father bought a plane for the family business, and aviation in the Case family became as contagious as measles.

For Christmas, Sandy's father-in-law gave the young Cases a Cessna Cardinal, shiny as the gilt-wrapped packages under their tree. On weekends Sandy and Tom hopscotched across the country to build up flying time, gain new certification. Sandy also had opportunities to copilot for the company plane, a turbo-prop. Her time was mounting.

During this period Sandy, her husband and children moved to North Carolina, where she continued her instruction. Being a wife and mother, running a home, hiring sitters and finding time to fly and study required sacrifices and a juggling of schedules. But she persisted. She earned her multi-engine, instrument and flight instructor ratings.

At last Sandy was ready to begin her career!

She applied to all the charter and commuter operations in the area but could secure only part-time work. To her consternation, the same companies hired other full-time pilots with far fewer credentials but with the one qualification that counted most— they were males.

"It was totally unfair and it stunned me," Sandy says. "But I could live with it then. I didn't have to work for a living, and part-time jobs gave me more time with my family."

So she took the jobs they offered, instructing and cargo carrying. Instructing, she reasoned, was an excellent way to build more air time, and the cargo flights would offer a new experience.

As the months raced by, Sandy became aware of a strange but persistent pattern in assignments. All the dreary night cargo flights were allotted to only two pilots—Sandy and a black male.

Even worse, the flights were scheduled in very old, small aircrafts. The choice daylight runs, complete with passengers and new, comfortable planes, all went to their male, white counterparts, regardless of seniority with the company.

Sandy, however, was being noticed as a competent flier. As she communicated with flight service stations, towers and facilities along the way, she began hearing rumors that the Federal Aviation Administration would soon start employing women in key positions. The FAA employees had been instructed to suggest qualified applicants. They recommended Sandy.

Sandy viewed the news as an incredible opportunity, a major breakthrough for women. Her marriage was floundering; she feared a divorce was in the offing. She needed a full-time job. Although a job with the FAA would not be a flying job, a position as weather briefer or air traffic controller would keep her in touch with aviation. And the pay would be a vast improvement.

With her hopes at high altitude, Sandy made the 500-mile-round trip for an interview at regional headquarters. Her enthusiasm soon plummeted.

These were the early days of transition in 1974 when women were just beginning to move into jobs traditionally held by men. People doing the hiring were often unsure, afraid to take the first step, intimidated by new federal laws and restrictions. Today the FAA offers the country's best opportunities for women in aviation. When Sandy applied, this was not the case.

"I was shuffled from the seventh floor to the first floor and back to the seventh," Sandy recalls. "Then the Civil Liberties people called me in for a conference. I left with no commitment."

A few months later she was called back to the same office for two major interviews. This, like the previous trip, would be at her own expense. But expense was a minor consideration. Surely they'd not request a return visit unless they had something definite in mind for Sandy.

It was Friday, a sparkling day full of promise. Sandy sat in the first interviewer's office calmly and confidently answering his questions, filling him in on her qualifications, her training and experience.

"Well," he said, tossing his pencil onto the desk, "I'm sure glad *my* two daughters chose nursing."

Dismissing Sandy, he added, "We've already hired two black men, so we're complying with the law."

Dejected, Sandy made her way to the office of the second interviewer but found he had gone to lunch. She was instructed to have lunch, then return at 1 P.M. She did. But when she arrived, a few minutes before the appointed hour, she learned the interviewer had left town but would see her on Monday if she'd stay over.

"I couldn't afford to make a scene," Sandy says. "They'd interpret that as a personality disorder. I didn't have money to spend the weekend, and I had to get back to my children. So I just had to grin and bear it."

She returned home.

In the following months Sandy's marriage did fall apart, just as she anticipated. Looking back, and with some regret, she realizes that her ambition was an underlying cause. If she had been content to remain wife and mother with no outside involvement, or if she had postponed her career until the children were grown, her marriage may have survived. But aviation careers must start early. To have postponed it would have dashed forever her chances to become a full-time career pilot.

Even so, her time was already running out. If Sandy didn't make it as a pilot by the time she turned thirty-five, she'd have to choose another career. She had one year left.

"I was scared," she recalls. "All my life I had had someone to take care of me—first my father, then my husband. I wasn't sure I could handle it alone."

In her anxiety Sandy considered a second marriage. But the possibility of another failure frightened her even more.

One thing she knew for sure. She couldn't manage on the part-time pay she was earning. She requested full-time employment again, and again was declined. A full-time position would involve flying charters, she was told. And charter pilots must load and unload their passengers' luggage, a job unfit for a woman.

So Sandy took out a bank loan and went knocking on doors. "I think I covered the entire South," she says. "Most of the time I couldn't even get an interview."

A friend told her of a pilot position open with a bank in a neighboring town. Her friend telephoned the president, outlined Sandy's qualifications, listing her certifications.

"Sounds like just the pilot I've been looking for," the president replied, his voice ringing with anticipation. "Send him right over."

"Uh," stammered the friend. "The *he* is a *she*."

"Forget it," the president snapped. "My wife would kill me."

So instead of hiring Sandra Case, qualified pilot and instructor, the bank official hired one of Sandy's students—a male.

By then Sandy had chalked up 2,000 hours in the sky and was eligible for commercial aviation. She applied at an airline headquartered in the South.

"Do you have any children?" the personnel manager asked.

"Yes," Sandy replied, wondering what this had to do with her qualifications.

"Then I'd suggest you stay home and take care of them."

Finally Sandy landed a job with a fixed base operation (charter and flight school) in Myrtle Beach, South Carolina. With joy and a sense of relief, she packed up her children, Wendy and Robert, their dog, cat and two horses, and set out for her new home. There, on the shores of the Atlantic Ocean, Sandy reported for work at her first full-time job in the career of her choice.

The job, however, turned out to be more than full time. The flight school was falling apart because the pilots already employed—all men—detested the mounds of paper work required. So Sandy, in addition to flying charter, took over the school.

"Suddenly I found myself working seven days a week and hiring an off-duty crew member to stay with my children," Sandy says. "The work was great. I was making a livable wage. But being away from Wendy and Robert every day was a constant worry."

Sandy, however, was carving out for herself a niche in aviation. Now she was flying daylight schedules, with sunlight glinting off the wings of late-model planes. In tow was a host of interesting passengers instead of the usual cargo of brown boxes and caged animals.

One of the company's regular charters was Senator Strom Thurmond, a South Carolina senator who maintained a resort home on a nearby beach. Sandy's first encounter with the well-known legislator was on a flight to a special ceremony on the lawn of a historic plantation in south Georgia. For the trip Sandy was assigned a small twin-engine aircraft requiring only one pilot. The senator, however, was accustomed to a plane with two pilots, both of them male.

"Where's the other pilot?" he asked, surprised, when he arrived at the airport.

The trip was filled with other surprises. Aloft, they soon left sunshine behind as they sailed into murky skies. Bouncing and bobbing in turbulence, Sandy wondered if her famous passenger feared for his life at the hands of a lone pilot—and a woman at that. To make matters worse, the airport in Georgia was shrouded in fog.

"I had to shoot an approach 'down to minimums'," Sandy says. "But we made a smooth landing, thank heavens, right on target."

The senator was so impressed he invited her to the celebration and introduced her from the podium.

Other charters proved even more adventuresome—but not as joyous. One client was a chronic alcoholic who regularly, while intoxicated, had to be rushed to various sanitoriums around the country. "Believe me, I always took a copilot for those flights," Sandy says. "He tried all sorts of tricks, anything from pulling at the hatch door to trying to jump out of a window." One time her passenger alighted, arranged another charter and arrived back home before Sandy returned to her home base!

Meanwhile Sandy's flying hours ascended and her credibility soared. She was rapidly approaching 5,000 hours, a magic number that would make her attractive to a major passenger airline. That would mean higher pay and fringe benefits, both of which she sorely needed. At thirty-five she was borderline but still acceptable if she moved quickly. She requested applications.

At the same time, a northern corporation made arrangements to base its new Beech Kingair at Sandy's airport and hired Sandy's employer to crew it. When the corporation executive was told that the major pilot would be a woman, his comment was not the familiar "Forget it" or "Does she have children?". Instead he asked only one question.

"Can she fly?"

Sandy flew to Wichita, Kansas, for ground school and was certified in the new Kingair. It proved a turning point.

At the school an attractive man asked her out to dinner. The school was large and there were many invitations for evenings out. But Sandy refused them all. The training was intensive and concentrated. She wanted to score well. So she stayed in and studied.

"Well, how about breakfast?" the man asked. "I'm with General Motors. Maybe you'd like to come to work for us."

Over eggs and coffee Sandy learned her host was in charge of ground operations for General Motors Corporation. He couldn't hire Sandy but wanted to recommend her to headquarters in Detroit, Michigan.

"It had been kind of a problem," Sandy explains. "General Motors was ready to hire a woman pilot but hadn't been able to find one who could handle the social situation."

The "social situation," according to Sandy, consisted mainly

Sandra before boarding Convair (photo by Wendy Case)

of convincing veteran male pilots with thirty-years experience that she was capable of handling their cockpit rather than merely decorating it.

"I tried not to get too excited at that breakfast meeting," Sandy recalls. "I loved the job I had. We were all like one big family. But I really needed shorter hours, more pay and security for my children. GM offered medical and dental insurance, retirement programs, paid vacations and a big raise in pay—the whole bit."

On the day she flew to Detroit for the interview, this time at GM's expense, her coworkers sent her off with a volley of "Go get 'em, Sandy!".

She did.

Applying at the same time was another pilot, a young male. "A real Listerine kid," Sandy says. "You know the type—neat, military haircut, three-piece suit, shiny shoes. He had flown the Sabre in the military, was an air traffic controller. He had every-

Sandra filing flight plans (photo by Wendy Case)

thing. I just knew he'd get the job, and I'd be sent scurrying back home to my seven-day-a-week work again.''

But she was wrong. They were both hired.

Now Sandy is a choice candidate for lots of jobs, and she gets offers. "When I took my routine tests with the FAA this past year, they tried to hire me," she says. "I told them they were eight years too late."

The advantages of her position as corporation pilot are many. GM has a reputation for keeping its fleet in tip-top condition, scheduling flights to give pilots plenty of between-trip rest. Unlike most corporations, her company has never laid off a pilot despite the many economic recessions in the automotive industry.

"With most corporations the pilot is always the first to go," Sandy says. "But GM keeps all theirs and uses the copilots as the third crew member in the back serving coffee. That's a whole lot better than being laid off."

Like the airlines, GM operates on a tenure (seniority) basis.

Sandra with one of her horses (photo by Wendy Case)

Sandy is now third in line to become captain, the first woman in the company's history to bear that title. As captain, she will be given a new automobile every three months on the Product Evaluation Program. She can select any GM car she wants. And she'll get a large increase in salary.

When her flight is not overnight, Sandy stays in a "day room" reserved at a motel near the airport. There she fills in her expense reports, reads, studies her flying manuals, shampoos her hair or suns by the pool—always within reach of the telephone and a possible change in instructions. She, like others in aviation, enjoys the erratic hours, the intrigue of not knowing where she'll fly tomorrow or how long the trip will last.

Sandy and her children live in a suburb of Detroit on the banks of the Huron River. They swim, canoe, cook out and ride horseback. On vacation Sandy zooms off to the Great Smokies for some hearty backpacking or to the Atlantic coast for sun and surfing.

Sandra relaxes with her guitar (photo by Wendy Case)

At other times, when she has a few consecutive days off, she pokes through the mountains of North Carolina, often on her trips to the hairdresser, in search of antiques. At home she relaxes with her guitar or a book.

Sandy averages ten to fifteen flying hours per week, to which is added her ground time. The ground time includes preflight and postflight filings, layovers and training sessions. Four days a month she remains "on call" around the clock for flights without notice. Her schedules are posted daily at 3 P.M. Monday through Thursday. On Friday the weekend schedule, extending through Monday, is posted.

Although her daughter, Wendy, would rather become a cartoonist, Sandy feels aviation is a choice field for the young woman who will persevere. She must begin early, Sandy says, build up air time as rapidly as possible, then go knocking on doors. Sandy also feels the military offers the very best in training.

"And despite my own experience, the FAA now provides excellent opportunities for women," she adds.

As for herself, Sandy will have the option of retiring at age fifty-five, and she plans to exercise that option. But she doesn't ever expect to give up flying.

"The other morning we were watching through the cockpit window as the sun popped up over the horizon," Sandy says. "No two sunrises are the same, but they're all magnificent. The other pilot commented that he sure hated getting up in the dark for these early morning trips, but there's nothing like being airborne when the sun rises."

Ditto Sandy Case.

MARY ELLEN KRAUS

□ ## AIR TRAFFIC □
CONTROLLER

MARY ELLEN KRAUS adjusted her headset.

"Wind one three zero degrees at one six," she spoke distinctly into the mouthpiece, the words spaced evenly on one unvaried pitch. "Altimeter two niner niner eight. Time zero two three one."

Through the receiving speaker came the sharp, clear voice of the answering pilot.

"Cessna three niner two one bravo, roger. Thank you . . . *sir.*"

The other controllers on duty chuckled in chorus, but Mary hardly noticed the "sir." Her voice is low, articulate and delightfully mellow. It's no new thing to be mistaken for a man on the radio transmitter or even on the telephone at home.

But Mary considers her voice an asset. "Lots of pilots aren't yet accustomed to hearing a woman in the control tower," she says. "Sometimes they even ask for a man."

One time in particular her voice provided the solution to a gripping situation. She was in the United States Navy then, serving a three-year tour as air controller at the Jacksonville, Florida, Naval Air Station. From a clear sky and forty-five miles at sea, an A-4 single-seat jet signalled trouble.

"Navy Jax, I've got a rough engine," the Navy pilot cried. "Request priority for emergency landing."

"Eight two four charlie," Mary's supervisor, a woman, radioed. "Proceed to . . ."

"Get me a man," the pilot snapped. "I'm in trouble. Nothing against *women,* you know. But I need to hear a *man's* voice."

The supervisor took a quick look at the only male on duty, a trainee much too inexperienced for an emergency. She handed the microphone to Mary.

Doubts raced through Mary's mind. Would it work? The pilot had to believe in that tower, that voice on the microphone. If he suspected, even for a second, that the voice belonged to a woman, it would unnerve him, shake his confidence. Playing tricks has no part in the serious business of controlling air traffic, but this was an emergency. She'd have to take the chance.

"Eight two four charlie, can you hear me?"

"Roger, *sir.*" His voice was edged with relief.

While the tower went into emergency procedures, alerting the fire station, clearing the runway, Mary took over tower-to-pilot transmission. Cooly, calmly she stilled the pilot's fears, guided him to a safe landing.

Once he was on the ground, the pilot telephoned the tower that he'd like to meet the man who saved his life, a request pilots often make after an emergency landing.

The "man" the pilot met was Mary—trim, attractive and very feminine.

"He couldn't believe it," Mary says. "His face was red, and he was mighty embarrassed. He said that's the last time he'd ever have a woman removed from the microphone."

Though her voice is pitched low, there's nothing about Mary's appearance or attitude that would dub her as anything other than female. She admits she was a tomboy while growing up. She scaled trees, built model planes and autos, played team sports. But that was because boys could do so many more fun things, she says.

It wasn't the pitch of Mary's voice, however, that catapulted her into a landmark position in aviation at the age of thirty-two. Instead it was her quiet ability, her self-discipline and dedication that captured the attention of the Federal Aviation Administration. In 1977, only six years after she joined its ranks, she was zipped off to Winston-Salem, North Carolina, to become the first female tower chief in the eight-state Southern region, one of only three female chiefs in the nation.

"That was the most terrifying moment of my career," Mary

says. "On the flight to North Carolina I grew more and more apprehensive—just plain scared, honestly speaking."

Her biggest fear was acceptance. As chief, she'd be the boss for the first time. Under her supervision would be two team supervisors and nine controllers—all men.

Further, hundreds of lives every day were dependent upon how the tower controlled air traffic in the area. And she'd be responsible.

"Finally I just had a good talk with myself," Mary recalls. "I told myself I'd just have to be *me* and tackle the job to the best of my ability."

Which is exactly what she did. Her acceptance as boss was not immediate. There were problems and mistakes in relationships. Mary made a few of her own. Working under stress demands a boss who can mediate, counsel and advise. Mary, accustomed to making quick decisions when directing air traffic, had to learn to listen patiently and intently to both sides of an issue before making a judgment. For Mary it was a growth experience, and she's grateful. When she was promoted from Winston-Salem two years later to the regional office in Atlanta, Georgia, her staff sent her off with sad farewells.

Mary's ability to handle air traffic, both routine and emergency, stems from two things, she says. A native belief in herself. And her extensive training.

After finishing high school in Minneapolis, Minnesota, Mary entered Washington University in St. Louis, Missouri. With more than two years of undergraduate work to her credit, she still felt uncertain about her major and her future. So she left school and spent a brief period in the business world. But a routine desk job was not the challenge she needed. Searching for new horizons, she joined the Navy.

When she finished basic training, Mary was offered three career fields in which to specialize—secretarial, medical assistant or air traffic control. She chose the third.

In what she terms a "rough thirteen weeks of school" she learned the essentials—navigational aids, chart reading, identification of aircraft, basics of radar and the actual air traffic control procedures. Upon graduation she was stationed in Jacksonville.

But that was only the beginning. Each time she graduated from one school, another waited. At the Naval Air Station she trained in control tower operations and radar approach procedures, learning each position so thoroughly it became as familiar as her own living quarters.

During this period, she handled the longest emergency in her career to date, a 2 $1/2$-hour battle against time, fuel and fate. Interestingly, it involved a private craft rather than a Navy plane.

She was on midwatch (midnight until 8 A.M.), a shift when only a skeleton crew was needed. Except for her supervisor, Mary was alone. She was working the radar position, her eyes glued to the luminous sweep on the dark green radarscope, when she picked up on her headset a very weak "Mayday, mayday, mayday," an international distress signal.

The pilot was flying a twin-engine Beechcraft, a charter flight from Miami, Florida, to Charleston, South Carolina, with one passenger aboard. He was lost. His gauge showed only forty-five minutes of fuel remaining.

Mary scanned the radar screen, which covered a sixty-mile radius. But the Beechcraft didn't show.

Feeling he must be flying far out over the ocean, lured off course by lighted boats, she turned him west toward land. Then she went into emergency procedures. She scrambled from the nearby international airport two fighter interceptors to fly out over the Atlantic Ocean far enough to pick up the lost pilot on their airborne radar. She activated the Coast Guard, setting in motion boats and airplanes capable of sea landings.

"Then I talked, talked, talked," Mary says. "I knew if I lost him on the radio I probably wouldn't be able to pick up his frequency again. The pilot was terrified. If I'd stop talking for even a minute, he'd shout 'Where are you, where are you?'."

With an eye on the clock, she coaxed him shoreward. She asked his name, the name of his passenger, what he saw when he looked out of the cockpit window.

"Every time he panicked I'd give him something to do," Mary says. "I'd tell him to look for the Coast Guard boats. I'd ask if he could spot the radar planes yet."

And the digital clock in the radar room quietly stole his time.

At thirty minutes Mary's supervisor began reviewing ditching

Mary Ellen in U.S. Navy

procedures with the pilot. At forty-five minutes Mary breathed a sigh of relief.

"You're still up there," she told the pilot, "so your fuel gauge may be incorrect. Stick with us. We'll get you in."

But how long did they have?

To relieve the stress, Mary alternated positions with her supervisor, gulped coffee, smoked cigarettes.

In time the airborne radar picked him up, but he was still many miles from shore. The interceptors moved into a circular pattern around the Beechcraft, marking the spot for the Coast Guard planes to land at sea.

Back at the controls Mary waited for what seemed inevitable —the splashdown, the frantic efforts of sea rescue against the black of night. But they never happened.

Two-and-a-half hours after the first weak "Mayday," Mary directed the pilot to a safe landing at the naval airport.

"He couldn't have had more than a teacup of fuel left," Mary says with a slight shudder. "An extremely lucky man!"

To Mary and other controllers, the tower becomes the center of her life. It can be situated atop the passenger terminal at the airport, but more often it occupies a separate structure some distance away. From a squat base building, housing the offices and the Terminal Radar Control room (TRACON), the tower rises straight as a pine to a glass-encased cab at its pinnacle.

Controllers inside the cab handle all ground traffic, all incoming and outgoing aircraft within a radius of five to ten miles. Then they "hand off" the departing plane to the radar controller in the TRACON below, who works the plane up to its assigned altitude. The TRACON controller tracks the plane for a sixty-mile radius, meanwhile "handing it off" to an en route facility.

From the cab the controller looks out upon a sky that seems alive with planes. At a distance they look like shimmering, nervous birds. She wears a headset and constantly either sends or receives messages. There is a chair at her station, but usually the controller is on her feet, pacing a small circle. Her eyes dart from plane to ground, then to the elevated radar screen above her head to verify what she sees. On hazy days she uses binoculars to determine a craft's numbers and colors.

The safety of the plane is the controller's prime concern. He

Mary Ellen receives the American Spirit Honor Medal in U.S. Navy.

or she is dealing with life and death, and the tension mounts. The standard joke among controllers is that "we're working on our ulcers today." They clown among themselves, share "inside" jokes and constantly tease. Usually one person bears the brunt of the day's jokes, then the next day it's another. The controller must learn to take it when it's her turn to be teased. It's all part of the job, a necessary component to relieve stress.

The cab is equipped with a sink, the makings for coffee, ash trays. Controllers take frequent breaks and rotate positions to prevent a tension buildup. If the position becomes especially stressful, the controller can request a break or a temporary switch to another station. It's his or her duty to remain alert, watchful, prepared for quick decisions.

By contrast, the radar room downstairs is called the "bear pit." It is windowless, quieter than the cab and darkened like a movie theater. The room is permeated with a constant low hum of equipment and the soft voices of the controllers on duty.

Radar screens are lined against the walls like a display of television sets. The controller sits before a dark green screen glowing with a myriad of yellow lines, blips (airplane targets) and numbers, like a sky of blinking stars. Every light gives some vital piece of information, anything from time and weather to the call number of each plane in the area. Each controller is assigned a segment of sky she is responsible for, her own slice of pie.

Because radar equipment can be damaged by liquids, no coffee is allowed inside the TRACON. But, as in the cab, controllers can smoke on the job, take frequent breaks and rotate positions. Controllers in both the cab and the radar room are required to make regular shift changes within the course of a week, since some shifts are more stressful than others.

"It's such an intricate process," Mary explains. "All you need to create a serious problem is for one cog in the wheel to break, one pilot to make a left turn instead of a right turn. The controller is constantly thinking, planning ahead, trying to read the pilot's mind."

During hectic moments, "rough words start flying all over the room." Now that there are women in the towers, the men often apologize for their language.

"But I tell them to just forget it," Mary says. "I don't use those words myself, but I have other ways of dealing with tension. I can go out afterwards and have a good cry. Men won't do that.''

Because job stress spills over into home life, the divorce rate among controllers is high. And the life expectancy of one who has worked the boards all his adult life drops to an alarming fifty-two to fifty-five years. These records are based on men, however, since women have not been in the tower long enough to become statistics. Most controllers retire after twenty years or request a transfer to a desk job.

After a shift Mary, who now lives in the Washington, D.C., area, unwinds by going outdoors. A wide expanse of space offers a sharp contrast to the confinement of the tower. She swims, plays tennis at her condominium, rides horses, shoots rifles. Because she works erratic and ever-changing hours, Mary finds time to pursue both daytime and nighttime interests. She becomes in-

Mary Ellen in cab of control tower (photo by John Huber, courtesy Federal Aviation Administration)

volved in the community wherever she lives, either in Scout work or backstage jobs with the amateur theater.

The one inherent problem with Mary's career is a big one, however. Loneliness. Mary is basically shy. She's reticent about initiating social contacts—a trip to the beach, a shopping spree, a movie or a dinner out. Since she works a shifting schedule, her friends can never be sure when she is available.

"So I end up doing a lot of things alone, or just staying home to read," she says.

Marriage at the moment is not in the offing, although she doesn't exclude the possibility. "But it would have to be a very special person and a meeting of the minds," she says. "Someone in aviation who would understand my weird hours and my dedication to my job."

Controllers, as others in the field of aviation, speak in a vernacular of their own. Because of this common bond, they tend to mingle with and marry others in the same profession.

"My mother says she hasn't heard me say Yes, No or What for years," Mary says. Mary instead uses the vernacular—Affirmative for Yes, Negative for No and Say Again for What.

Despite its drawbacks, Mary finds being an air traffic controller an excellent career for women. "Once your peers learn you can do the job as well as a man, that you're not asking for special privileges because you're a woman, that you'll do just as many dirty chores as they do—like emptying trash cans or making coffee when it's your turn—then you're accepted," Mary says.

Mary admits she still likes it when a man opens the door or performs some other chivalrous act for her. It makes her feel special. But she doesn't expect it on the job.

"I sometimes have to struggle with a desire to be protected by someone bigger and stronger," she says. "But if I have to, I'll stand up and fight for myself."

Mary no longer works inside the tower, although she is still in the Air Traffic Control Division. To advance with the Federal Aviation Administration she must accept promotions and transfers in a normal career progression. From the tower in Winston-Salem she became a planning specialist in the regional office in Atlanta. In this position she coordinated the expansion plans for thirty-two air traffic control facilities in the two Carolinas, an overwhelming job with demanding responsibilities. From Atlanta she progressed to the national headquarters in Washington, D.C., where she works on the National Tower Relocation Program, along with several other national programs. Her rise has been rapid, her performance outstanding.

Would she like to fly her career all the way to the top, to become chief of the largest airport in the world? For Mary there's just one word to supply the answer.

Affirmative.

JOYCE CARPENTER MYERS

□ AERONAUTICAL □ ENGINEER

JOYCE FELT THE TIP of the stethoscope, cold as an ice cube, slide across her chest.

"What are you going to be when you grow up?" the pediatrician asked.

"An astronaut." No hesitation.

"Good for you." Dr. Leila Denmark slipped the ear plugs down around her neck and looked her young patient squarely in the face. "Stick with it."

The visit to the pediatrician was routine. There was nothing wrong with Joyce, no disease a stethoscope or thermometer would reveal. But the twelve-year-old did suffer a common malady of the 1960s—the Walter Cronkite Syndrome.

"Every time there was a space flight I'd skip school and sit glued to the television," Joyce says, a slow smile creeping across her face. "I still haven't recovered."

Joyce had always admired her female pediatrician. Not only had she become one of the best doctors around, but Dr. Denmark was codeveloper of the measles vaccine. And she, like Joyce, dared for heights the average person wouldn't try. Although she was more than seventy years old, Dr. Denmark spent her summer vacations scaling mountains. Her "prescription" to Joyce to "stick with it" would be easy to stomach.

A few years later Joyce received another "prescription" regarding her chosen career, this time from the National Aeronautics and Space Administration (NASA). Joyce had been obsessed with the space program since age nine, when Alan Shepherd carved out that first thin slice of sky. During summers, she coerced her family into pilgrimages to Cape Kennedy long before there were guided tours or anything to see but a few markers and a Holiday Inn. She lined her bedroom with science fiction, peppered her walls with color photos of astronauts and space ships. And planned.

In high school Joyce wrote a carefully worded letter to NASA asking how she could become an astronaut, a moot question since there were no women in the program. NASA's reply: "Get an engineering degree." In a postscript the agency added: "And it wouldn't hurt to learn to fly."

NASA's "prescription" was a massive dose for a teenager who couldn't yet drive a car and was several years away from college. But Joyce set out to comply.

She dealt with the postscript first. She joined the Civil Air Patrol unit at her high school in Dunwoody, Georgia. In a well-coordinated cadet program she sampled both civilian and military

Joyce (left) as Civil Air Patrol staff sergeant at summer encampment in Montgomery, Alabama, at age seventeen. (With her are lifelong friends Virginia Kluge (center) and Lynn Smith (right).

Joyce (center) as Civil Air Patrol cadet in high school

aviation. She wore an Air Force uniform, marched, drilled, took field trips.

And learned to fly!

"I actually got my pilot's license before I got a driver's license," Joyce says. "It was a whole new skill, learning to operate a vehicle. A lot to soak up."

That summer she won a scholarship to the Civil Air Patrol Flying Encampment at Oklahoma State University and continued her pilot training. She grew comfortable at the controls—and good at flying. NASA's postscript had been accommodated.

Now for the big one!

The closest engineering school, and one of the best in the nation, was Georgia Institute of Technology (Georgia Tech), her father's alma mater, in nearby Atlanta. But Georgia Tech was a rigid stronghold for 7,000 topflight students—almost all men.

When she discussed her choice with her guidance counsellors at school, she received mixed messages. One, a woman, said engineering was too hard for a girl, that Joyce would grow discouraged. The other, a man, refused to believe she could be accepted at Georgia Tech.

Flanked by photographs of astronauts John Glenn and Scott Carpenter, Joyce spent her afternoons poring over math and science books, kept her grade levels high, sent in her application to Georgia Tech and haunted her mailbox for a reply.

Finally, just before graduation, the answer arrived. Joyce was accepted!

Wearing the famous Georgia Tech "rat cap" to signify she was a freshman, Joyce moved on campus in the fall of 1968. So did 249 other young women—but there was no woman's dormitory. For living quarters she could choose between the top floor of an old Young Men's Christian Association building or a small, cramped house. Joyce selected the house, which she shared with eleven other women. Although she had three roommates, Joyce, who calls herself a loner, grew close to none of them.

"Those were painful days at Tech," Joyce says. "We were all so busy trying to prove we were as good as the men that we didn't have time for friendships. We were all in different phases of engineering, had different schedules and hardly saw one another."

In class the male students scurried to the other side of the room, careful not to sit next to a coed for fear of losing stature. And the campus rocked with endless coed jokes.

That first year Joyce kept the prospect of the NASA-inspired degree in the forefront and pushed all thoughts of flying to the back of her mind. But it concerned her that she was not building any pilot-in-command time or improving her flying skills. In her sophomore year she joined the Tech Flying Club.

And found a special bonus. His name was Mike Myers. Mike was in aerospace engineering, too. Three years later they were married.

Meanwhile, Joyce and Mike pooled their resources and shared the rental of a plane. In June 1972 Joyce finally held in her hands the two specifics NASA had prescribed—an engineering degree and an up-to-date pilot's license. She rushed to inform the agency.

But NASA wasn't impressed. They still were not accepting women into the program. She'd have to be patient.

Joyce with coworker Kay Cornelius, also an engineer (photo courtesy Lockheed-Georgia Co.)

Joyce stepped back and took a long, lingering look at her life to date. Her engineering degree was gained primarily as a springboard into space. Now she'd have to use it to earn her livelihood —and she wasn't even sure she wanted engineering as a career.

To make matters worse, the field was crowded; there were very few jobs available, and the companies with openings much preferred men. But even Mike couldn't find a job.

They learned of an opening in the Peace Corps and applied. But the requisition called for only one engineer with a graduate degree and five years experience—all for eleven cents an hour.

In despair Joyce and Mike combed the job markets for other types of employment.

"The swing of the time was away from materialism," Joyce says. "So Mike and I thought maybe we'd like to farm, live off the land."

They set out for Missouri in a 1965 Dodge Dart to join some friends, Kathy and Paul Brautigam, who were trying their hands at farming. On a small piece of land tucked at the end of a rural road, Joyce and Mike became "Farmer Brown and his good wife." Joyce was in charge of the barnyard. She fed chickens, gathered and sorted eggs, helped select new stock. Mike worked on farm equipment and house repairs.

"Actually we were poor city kids playing like farmers," Joyce says. "But we liked the feel of the land. We decided we wanted a farm of our own."

So they scooted back to Georgia to earn the down payment. Moving in with Joyce's parents, they felt the quickest way to make money was to drive a tractor-trailer. They both enrolled in a driving school.

Joyce, along with the men in her class, learned to swing a tremendous rig around a parking lot behind the school, took short evening road trips, studied and got her Class-Five commercial license.

And faced another disappointment.

Joyce and Mike were both twenty-three years old. Insurance regulations for commercial drivers under twenty-five are stringent. Freight companies looked with skepticism at their youth, particularly for assignments to drive together. But Joyce and Mike persisted and finally landed a contract.

Together they hauled an eighteen-wheeler, loaded with margarine, to Portland, Oregon. From Portland they rolled to Indio, California, for a load of fruit destined for New England. In Brooklyn they picked up a cargo of plastic for Los Angeles. They drove straight through, day and night, taking turns sleeping in the bunk in back, stopping at truck stops for showers and food.

Their life was far different from the way they had planned it at Georgia Tech. Yet they liked what they were learning—to live unencumbered, work hard, be flexible. But in Tucson, Arizona, their driving career ground to a resounding halt. The company they were driving for didn't own the truck, hadn't paid rental, and their rig was repossessed on the spot.

"It was a devastating experience," Joyce says, "but it turned us around in our thinking."

Joyce and Mike hitched a ride with another truck driver back to Georgia. En route they reassessed their future. While whirling down interstates, their cargo far behind in Arizona, they decided to try again for jobs in engineering. The nation's economy was beginning an upswing after the recession of 1974. The job market looked promising.

It was. In January 1975 both Joyce and Mike signed on as aerospace engineers with one of the world's largest airplane manufacturers, Lockheed-Georgia Company in Marietta. Joyce was assigned to structural strength, Mike to structural dynamics.

Dressing for work that first day, Joyce felt a wave of self-doubt wash over her. She was three years out of college with no professional experience to her credit. The days of feeding chickens in Missouri, the nights of roaring down highways at the wheel of a tractor-trailer had enhanced her personal growth. But they certainly did nothing for her engineering ability. She had brushed up on her textbooks, but how much had she forgotten? What new concepts must she learn? And to make matters worse, the old saw that "to succeed in engineering, a woman must be twice as good as a man" rose like a ghost to haunt her.

"I was the first woman Lockheed-Georgia had hired in years," Joyce says. "I was scared."

The male engineers in her office, while courteous and kind, were also skeptical. They watched her closely, guardedly, during those initial weeks for the first telltale signal of her inadequacy.

Joyce, meanwhile, subdued her fears with hard work, with a willingness to learn and cooperate. It worked.

"Fortunately engineers pride themselves on professionalism," she explains. "If you do your job well they eventually forget your sex, quit asking you to make coffee and accept you as a pro."

Joyce was assigned to a team of six. The other five members are men. Their project is titled Durability and Damage Tolerance Assessment on a C-141 jet transport used by the U.S. Air Force. In the late 1960s, Lockheed-Georgia removed a shiny C-141 from the assembly line for use in a simulated flight test. The plane sits like a mammoth bird inside a hangar at the company complex where Joyce works. Although the plane is hemmed in by walls and a ceiling, its cargo compartment is pressurized and its wings are jacked up and down as if flying. This simulation takes place in a constant twenty-four-hour-a-day syndrome and duplicates the stress incurred by actual flight. It's part of an intricate and vital testing system designed to insure airplane safety.

Joyce examines wing of C-141 in Lockheed-Georgia hangar. (photo courtesy Lockheed-Georgia Co.)

"The staff at Lockheed maintained a continual check, constantly examining the wings for any evidence of damage," Joyce says. "Sure enough, after the tenth year a crack appeared on one wing."

The crack was cut out and removed to the company laboratory, where a metallurgist examined it for markings. From these tests Joyce and her group determined the exact position of the plane at the time the damage occurred.

Joyce compiles the on-going data, which she calls "fractographic information," makes charts and graphs. Then she feeds the information into a computer, along with variables such as weather, cold, heat and wind, to predict the exact moment such damage is likely to occur, how to repair it and—most important of all—what to do to prevent it.

The C-141, the plane on which Joyce works, in flight (photo courtesy Lockheed-Georgia Co.)

Mike, meanwhile, works in a different area of the huge complex, and they seldom see each other during the day. They both resist any appearance of a Mike-and-Joyce duet on the job. Joyce is determined to be recognized for her own merits, to be "Joyce" rather than "Mike's wife."

Working for the same company in the same complex offers a contrast of reactions. Joyce finds that men are less apt to flirt— and she detests flirting. But both Joyce and Mike are highly motivated, career oriented and competitive. All three aspects are desirable for their professions but could easily cause a fracture in their relationship like the wing fracture on the C-141.

"Although we're not actually competing for the same job, we are competing in other areas of our professional lives," Joyce explains. "In salaries, company favors, assignments, trips. All these things constitute the measure of our success."

In engineering it's often necessary to move to other assignments, to other companies even, in order to grow professionally. Both Joyce and Mike feel positively toward Lockheed-Georgia, but Joyce fears if she must "push" for another assignment or a promotion it could reflect unfavorably on Mike.

An advancement could well require a move to another plant, a resettling in another state, a separation from Mike. They've faced the prospect.

"We're rather conventional," Joyce says. "We don't have one of those far-out marriages. On the other hand, we don't think our relationship hinges on one central location. We'd be just as married if I were in California or Alabama as we are living together here in Georgia."

The Myers don't plan to have children, so a central "nest" is not as important as it might be. Joyce would come home as often as possible, and they'd work at making their time together a special experience.

"We could even talk about our jobs then," Joyce says with a shy grin. It's something they avoid now.

Home for Joyce and Mike is an old farmhouse and six acres they are buying in a rural area northwest of Atlanta, near Marietta. They call their home Ahoghill, named for a town in Ireland. Currently they do only a little gardening, but someday they plan to stock the farm with cattle. Their jobs at Lockheed-Georgia are too demanding at the moment to add extras. Both Joyce and Mike have earned their Master of Science degrees in engineering while working, and both will probably continue their graduate studies.

Joyce and Mike share home duties but find, with one exception, that the chores follow the usual line of division. Mike does the fancy cooking! Armed with two thick French cookbooks, he prepares mouth-watering spinach souffles, beef Wellington and whole grain breads. To Mike cooking is creative. To Joyce it isn't. She prefers a microwave oven and easy meals. "It may be a cop-out, but it gets food on the table in a hurry," she says.

Joyce does the laundry and most of the housecleaning. She sews, does handicrafts, and the two of them spend any spare time remodeling their house or working on their automobiles (they still have the 1965 Dodge Dart that carried them to Missouri).

The one outside activity Joyce consistently makes room for is her work with the Society of Women Engineers. She's president of the Atlanta section and has been instrumental in establishing a program to encourage high school girls to consider a career in engineering.

"It would have been so much easier for me and my classmates if we had been given some exposure or encouragement when we were in high school," Joyce says. She still remembers with dismay her own discouraging counselling and realizes it could well have turned her away from both Georgia Tech and an engineering degree. She'd like to prevent a recurrence.

Under Joyce's leadership, the Atlanta Section of the Society of Women Engineers provides speakers to go into the school system and tell about engineering as a career for women. Her group also offers special incentives, such as cash awards or a Certificate of Merit, for engineering projects; and they're working toward establishing scholarships.

The most important qualities for engineering, Joyce says, are an aptitude for math and science, a predisposition to set things in order, to see things logically and to learn self-discipline.

"In high school there's no real reward for working hard except to get good grades," she says. "The average student doesn't translate that into preparation for a career. It takes self-discipline to become an engineer. The earlier you learn it, the better."

Her projects at Lockheed-Georgia, for instance, involve seven to nine months of steady work for each unit. Joyce must set goals daily, otherwise she'd reach midpoint and find herself with three-quarters of the assignment uncompleted. She charts her progress on a desk calendar.

Joyce was fortunate. She learned to think constructively at an early age. On Sundays her father always initiated a critique of their minister's sermon at lunch. Over fried chicken and mashed potatoes, Joyce, her brother and her parents dissected the pulpit message. First they explored the rhetoric. How skillful was his language? His delivery? Was it appropriate? Then they searched for the minister's primary lessons and next for underlying meanings.

"My father always led the discussion," Joyce recalls. If the minister used a new word, they had to look it up in the dictionary before they could dive into their strawberry shortcake.

Those early intellectual exercises probably turned Joyce toward a profession whose purpose is to find answers. Although her dream of becoming an astronaut has not been realized, Joyce feels challenged in a career that bears a kinship with aerospace. Her goal now is to become the very best engineer she possibly can be, to operate at her maximum potential, to scale mentally the mountains her pediatrician scaled physically.

She's well on her way.

ANN ORLITZKI SMETHURST

□   CAPTAIN   □
U.S. AIR FORCE

THE DESTRUCTION of the entire world could be triggered a few feet behind Ann Orlitzki Smethurst's back as she quietly goes about her job.

The power to activate retaliatory nuclear weapons is embodied in a massive console aboard an EC-135 airplane. As copilot, Ann plays a leading role in an on-going drama that is both awesome and assuring.

The drama is a mini-ceremony performed three times a day at Offutt Air Force Base, Nebraska, under tight security. There are no blaring bugles, whirring cameras or curious spectators. In fact, most residents of the United States don't even know of its existence. Yet their survival may depend upon this ritual.

It proceeds like this.

An operations officer and a communications officer, both carrying side arms, receive from the Strategic Air Command headquarters a sealed package containing top-secret authenticators and code information. Escorted by armed guards, the officers carry the package to a designated EC-135 and place the material inside a safe. The safe door is meticulously closed and double locked.

The two officers take over their stations inside, and the plane lifts off. On board a complete battle staff is assembled, including a general and ten highly trained specialists, five members of the communications personnel, plus the crew. As the plane gains altitude, it begins radio dialogue with a descending EC-135 now completing an eight-hour mission. A switch-over is inaugurated. The ascending craft at that moment assumes the weighty responsibility for the nation's nuclear defense for the next eight hours.

The program is based on the alarming knowledge that the Land Command Control, strategically placed far below in underground Nebraska, could be destroyed in the first forty minutes of a nuclear attack. By keeping a duplicate Air Command Control flying high in the sky, in immediate touch with every Strategic Air Command post in the world, the United States is prepared to defend itself.

The flying mission is called Operation Looking Glass. The first plane went aloft on February 3, 1961, and a plane with the same ability has been in the sky constantly since that date. Its battle staff does not have authority to activate nuclear weapons. Only the president of the United States has that power. The staff does, however, have the responsibility of carrying out that order should it come.

Ann is the only female pilot in Operation Looking Glass.

Ann happy at the controls (U.S. Air Force photo)

A year or so ago Ann had an experience with another type of looking glass, one more familiar than the military version. She was in California on a temporary assignment when she suddenly decided to buy a dress.

"I hadn't bought one in ages," she says, "hadn't even looked at styles."

Ann loves color. She weaves bright rugs, delightfully fingering each strand of yarn as she works it into a pattern. She decorates cakes with special tints. Ann also likes dressing up once in a while, wearing a cocktail dress and heels for special occasions. But mostly she's content to wear her single-color uniforms with a white shirt, or plain off-duty casual clothes.

That day she went shopping and selected a peach-colored gauze, a filmy, flowing sort of dress.

"I saw myself in the mirror down the hall and actually said out loud, 'Is that me?' I wasn't fat and I wasn't ugly," she says "Until that moment, that one tiny moment, I had always thought of myself as a beached whale."

Ann calls this her "California experience." It was a significant event in her life, and she refers to it with a sense of awakening. For the first time she was pleased with what she saw in the looking glass. Now with a philosophical bent and a dash of her keen humor, she says she can look in the mirror and say, "Such is life. You deal with what you've got." She may never think of herself as pretty (she is), or shapely (she is). But she's at least elevated her self-image to one she can accept with ease.

The fact is, Ann had never thought of herself much at all. She was always a doer, a perfectionist bent on accomplishments. While her classmates strolled home casually from school, Ann raced to the airport for flying lessons. When her peers tried new makeup or discussed boys, Ann practiced landings. While they spun the latest rock record, Ann was soloing to a melody of her own.

Ann did these things alone or under the watchful eye of her parents. She had no sisters or brothers, no close friends to ask, "Hey, how do I look with my hair fixed this way?" or "What should I wear to that?".

Because her mother is a beautiful woman, an artist and former Hollywood startlet, Ann's self-denial became more complex. To Ann, her mother had everything she lacked and could never have—

"big bosom, small waist, gorgeous face." Ann took delight in the beauty of her mother and ignored her own special appeal.

When Ann turned eleven, she already stood five feet, nine inches tall, her present height. She towered over her peers and felt like an Amazon. Her mother had to arrange an escort to the senior prom.

"I never even kissed a boy till I was seventeen," she says. "And to this day I've never really had a courtship." She married her husband, Richard Smethurst, after only four dates and "a lot of letters and long distance calls." On their dates they discussed lofty subjects —politics, patriotism, affairs of state—and went out to dinner Dutch treat. They learned to know and love one another more through letters than by being together.

After her "California experience" Ann can look back and realize the reason she didn't date in high school and college was because she

Ann and Rick leaving the chapel after their wedding (U.S. Air Force photo)

was involved in something foreign to her classmates. She was considered bright and academic. She constantly pushed herself to be best. She frightened them.

"And the irony is that I was pushing because I felt so inferior," she says. "With parents like mine—achievers—I felt I had to really accomplish something for myself."

Ann's father is an ex-Air Force colonel, a veteran of World War II and a dominant force in her life. He encouraged Ann to set goals, to achieve. At fifteen Ann took flying lessons near her home in Fairview, Massachusetts. At sixteen she soloed. As a sophomore at the University of Connecticut she became the first female cadet in its Air Force Reserve Officer Training Corps, one of only thirteen nationally, and received a full-tuition scholarship.

By the time she graduated in 1973 with a degree in both philosophy and physics, Ann had not only worked out a personal philosophy but a way of looking at life as a logical process. Her philosophy is what she calls "practical idealism." She keeps her eye trained on the ideal (to succeed) but works on it from her current level (by doing the best job at all times, whatever the job is). Her logic came into play when she chose a career. She wanted aviation because she loves to fly. She chose the military because she loves her country. It's that simple, that logical.

"I'm so patriotic you can't believe it," she says. "I even jump up and salute when they play the 'Star Spangled Banner' after the 'Late Show'."

At the time Ann made a career choice, the Air Force did not permit women to fly. But rumors were rampant that someday women would be admitted to pilot training. Ann was in that first class.

Shortly after Ann had her looking-glass experience, which enabled her to be comfortable with herself as a woman, she had another experience that enabled her to accept herself as an officer. She was a captain and already flying Operation Looking Glass when she was selected to attend a special three-month school at Maxwell Air Force Base in Alabama. The school was designed for "middle management" officers and consisted of long classroom hours and a series of demanding, sometimes grueling, outdoor projects.

Arriving on base, Ann found she was the only woman pilot among 625 classmates. Most of them had never seen a female Air Force pilot and certainly had never worked with one. They were suspicious. Rumors swept through the barracks that she was lesbian, that she hated men, detested flying. Some of her classmates accused her of having affairs with officers of higher rank. And one officer said he wanted to "have" her sexually because he had never "had" a pilot before.

When she was graded "excellent" on a crucial test, a classmate openly charged her with using "femininity" on the instructor.

"He kept making digs and nasty comments during class, censuring me for all sorts of things," Ann says. "A man in my position would have invited him out in the hall and put a couple of teeth down his throat." All Ann could do was confront him verbally.

"This is beneath you as a person," she told him after class. "Why don't you check your facts?"

For this outburst Ann was criticized, on record, as being "too defensive."

Ann planning a mission (U.S. Air Force photo)

The school was a trying experience, sometimes an ordeal, but Ann won the respect, and eventually the admiration, of the other officers. The emotional pain is still there, she says, but she worked through it and came out on top. Again she was able to upgrade her self-image. Now she feels totally comfortable with herself as an officer.

"I know I don't have to swear like an S.O.B. or smoke a cigar to be a good officer and leader," she says. "That image is totally irrelevant to the function. I'm not a 'woman officer' and a 'woman pilot.' I'm a woman who happens to be an officer and a pilot. I'm a captain, not a captainess. I like the feeling of equality. I'm glad the military uses the same title for both men and women."

But titles are not important to Ann. Nor does it matter to her what position a person holds. "As an individual, an astronaut is no better than a garbage collector," she says. "It's whether or not that person tries to be the best darned astronaut or the best darned garbage collector possible."

Ann took her first flight in a plane when she was three days old. At three years she was able to flip to her favorite pages in the encyclopedia. One page featured a picture of Charles Lindbergh. The other was a spread of brightly colored flags of every nation. She memorized the flags and their countries, the caption under Lindbergh's photograph. She spent hours making airplanes from scraps of paper.

By the time she reached grade school a poster of the seven original astronauts hung over her bed, and her favorite pastime was finding the moon and stars in her telescope. She adopted the binary and variable stars as her circle of friends. Alone with her telescope, she "talked" to the stars, told them her secrets.

As a teenager, Ann burned with a desire to penetrate the friendly sky at the other end of her telescope. Almost everything she remembers about her teen years involves trying to make it possible to fly among the stars. On the day she was scheduled to take her written examinations for a pilot's certificate, the high school principal refused to excuse her from classes. Flying could be of no practical value to a woman, the principal said. Ann would have to take zeros and make up the classes.

"The irony of that principal's remark makes me chuckle now," Ann says, now firmly anchored in a flying career both profitable and challenging.

After Ann joined the U.S. Air Force in 1973 and finished her basic training, she was sent to North Carolina State College for graduate studies in meteorology. While it wasn't flying, Ann found she loved working with the weather and dealing with the male pilots who were fortunate enough to be in the sky. At school she met her future husband, now a captain in the Corps of Engineers. They were married at Christmas, and Ann requested an assignment at Pope Air Force Base, only a few miles from Fort Bragg, North Carolina, where Rick was stationed.

Two years later she learned that the Air Force was offering its first program for women pilots. Encouraged by her husband, Ann applied and was accepted.

Ann learned early in her marriage that an advancement for either of them means months, sometimes years, of living apart. The military allows couples to file Form 90 requesting assignments at the

Husband Rick pins Wings on Ann on graduation day. (U.S. Air Force photo)

Ann poses with her plane. (U.S. Air Force photo)

same location. But for the Corps of Engineers to be stationed on or near an air base is something just this side of a miracle. For Ann and Rick this has happened again at Offutt Air Force Base, but this arrangement could be terminated at any time.

When Ann was in pilot training, Rick was based in Washington, D.C. The day she won her Wings he was preparing to go overseas. When she was in California, he was on his way to Korea. And in all their years of marriage they have shared only one vacation.

Pilot training and that first separation from Rick were the lowest points of Ann's career, probably the lowest points in her life. For some reason, which Ann does not fully understand, she felt inferior.

"I thought I'd flunk for sure," she says. "And even worse, I learned some horrible things about people. I learned about jealousy, about how people will tell you untrue things just to put you at a disadvantage."

During this one-year period, Ann saw her husband three times, one week each. "He didn't even have a telephone, so I couldn't just up and call him whenever I needed to," she says.

Ann again is facing a decision involving separation. She is in line to become a Commander, a lofty position that puts her in the cherished left seat of the plane and in charge of the crew. This would involve going away to school. If she goes now, she'll be the first woman commander in Operation Looking Glass. To Ann, that's no reason at all.

"There's no joy in being first," Ann says. "I don't even see it as a reason. All I want is to be the best I can possibly be. If I'm fiftieth down the line it doesn't matter a whit."

What does matter is being away from Rick. "I know we grow when we're apart," she says. "But I'm putting my marriage on the line. There's always the risk that the growth would be in opposite directions, that separation would drive a wedge in our relationship."

From her current vantage point, Ann says she would now postpone pilot training until her marriage was older, more mature. When she was in school at Maxwell Air Force Base, a young woman, an Air Force officer, told Ann she was considering applying for pilot training. "You have my condolences," Ann said and walked away, leaving a bewildered would-be pilot in her wake.

"I later went back and apologized," Ann says. "When she first made that comment, I just didn't have the heart to tell her what it's really like for a woman."

On the second encounter Ann told the woman to be sure she is as emotionally strong as possible, that her marriage is stable and that she is secure as an individual.

"She's married to a big blond guy; she's a little brunette with a pretty face, and they both grin a lot," Ann says. "A lot of the things they have together now will disappear. And much of the joy of being a pilot will be taken out of it by her male counterparts."

According to Ann, once you set on a course a little different from the norm you're never again the same person. "There's the reality of learning just how low people will stoop to compete. And the old double-whammy—learning not to be sensitive without giving up some of the joy of living," she says.

In pilot training a senior officer asked Ann if she realized how unique she is. "No," Ann replied, "and I don't want to know. If I ever do, then I've lost my innocence, and my reasons for doing things may be different. I don't want to change that."

Because she was in the first group of women Air Force pilots, Ann was pointed at, talked and written about. She was photographed for the media, invited to make speeches. Ann could have easily turned this into an ego trip.

Instead she asked that her interviews be cancelled, that her publicity be halted.

"I just didn't think of myself in those terms," she explains. "I'm simply one member of a team."

Young women often approach Ann with questions: "Is it hard to learn to fly?" "Do you think I could?" Their questions are tinged with their own self-doubt, and they add, "I don't have a mechanical mind, you know," or "I have a fear of heights."

Yet these are the same women who poke the vaccum cleaner back home to make it work, Ann says. And the ones who face the pains of childbirth with courage.

Ann has dealt with fear herself, many times over. Just before she was to solo in a T-38, she rushed to the bathroom with an upset stomach. "What are you afraid of?" someone asked. "Dying," Ann replied.

But she soloed.

"Women can bake a cake, but they think they can't be artists," Ann says. "They carve a roast, but they think they can't be surgeons. They manage a household, but they think they can't run a corporation. What I say to these women is to realize that the same skills can be used for other things. The skill has nothing to do with the role. The ability is independent of what we're doing. Separate the skill from the role, then expand that skill."

Although Ann does not think of herself as physically strong, she's never missed a day due to her menstrual cycle. She's run three miles when she'd rather curl up with a hot water bottle. She's flown aerobatics and pulled seven G's in a plane (a force equal to seven times the pull of gravity) on days when she felt "delicate." "I'm not always at my peak every day of the month," Ann says. "But then, neither are men."

Ann reviews the myriad of buttons she must know. (U.S. Air Force photo)

At Maxwell Air Force Base one of the leadership tests involved a large gravel pit charged with simulated explosive land mines, a tree stump, a rope and an axe. Ann's project was to get her seven men safely across the pit. It's a psychological study, an opportunity to show creative thinking, designed by Germany in World War II as a test for selecting commandos.

While Ann was leading her group through the pit, one man turned to the instructor and exclaimed, "God, that woman has the best ideas of anybody out here. I just wish we knew *when* to listen to her."

To Ann this translated into a lack of leadership on her part. "I should have been able to motivate that guy to listen to me *all* the time, to get him to see me as an officer instead of a woman." It's the same old story, she adds. A woman has to work *twice* as hard.

Women, according to Ann, can often be more creative even in gravel pits. "It's like fixing the dishwasher with a bobbie pin," she says. "Women are accustomed to thinking creatively."

As a further test, the civilian wives of some of the officers were asked to go to the same gravel pit. They went reluctantly, expecting to perform poorly. But they showed imagination, had a good measure of success and wanted to go out again.

When flying Operation Looking Glass, Ann is functional to the tips of her nerve endings. Every flight has to be acted out as if it were "the real thing," a response to a nuclear attack. Her plane flies random patterns, varying altitudes and a constantly changing schedule, all devised for security. Each mission lasts eight to nine hours. Ann flies fifteen missions each month.

"It gets boring flying around for nine hours, and we joke about it on the ground," Ann says. "But while we're aloft we're at tip-top all the time. You don't relax a minute. That's your mark of professionalism."

Some mornings Ann gets up at 4 A.M., finds snow piled against the windows of her suburban home and learns the chill factor has plummeted below zero. But she arrives at the air field well before flight time. While her ears feel like ice cubes and her breath makes small marshmallow circles in the air, Ann makes her preflight checks. It's her responsibility as copilot to determine that the plane is in perfect condition before takeoff.

Other days she's aloft when the inside thermometer soars to 120

Ann checks every nook and cranny before a flight. (U.S. Air Force photo)

degrees Fahrenheit and the air conditioner can't possibly keep the plane comfortable. But Ann stays "right up to alert."

When she's not flying, Ann works in the simulator, constantly upgrading her skills. She leads and participates in seminars and is now beginning flight instructing. Her working hours range from sixty to eighty hours a week. This year she spent seven months in school.

Ann would like to become a pilot astronaut. "Not a mission specialist," she explains, "but the pilot who actually flies the spacecraft." Were she chosen, the Air Force would grant her a leave. But the chances are slim. Becoming a pilot astronaut requires an abundance of hours in high-performance aircraft. The easiest and fastest way to log these hours is in a fighter. But women are not permitted to fly fighter planes.

Her second choice is to fly a C-9 in the Presidential Squadron out of the nation's capital. "This is the plane that flies the Secretary of State around," Ann says. "It's a beautiful plane with a very special crew. You're actually representing the United States to the rest of the world. I'd like to get involved in that."

To be pilot of a spacecraft or in the Presidential Squadron is to excel. Few are chosen. But Ann's philosophy holds that greatness is not destined, it's something you choose yourself. Then you make it work.

Ann will.

MARGARET RHEA SEDDON

□ ASTRONAUT □

IT WAS SHORTLY before 8 A.M. on January 16, 1978. Dr. Margaret Rhea Seddon was making her customary rounds up and down the brightly lighted corridors of the Veterans Hospital in Memphis, Tennessee.

As she moved from room to room, her sunny smile a welcomed sight to the patient in bed, Rhea (pronounced Ray) had no way of knowing that within a matter of minutes the course of her life would take off like a rocket. That within a few hours she'd be holding a press conference important enough to be flashed around the world. Or that within twenty-four hours she'd appear as a celebrity on the "Good Morning America" television show in New York City.

The sequence of events began when the paging device in the pocket of her white surgical jacket beeped her to the telephone.

"Are you still interested in becoming an astronaut?" the voice at the other end asked.

"Sure!"

"Then why don't you come on down here in July and we'll start your training?" George W. S. Abbey, director of Flight Operations at NASA-Johnson Space Center in Houston, Texas, said.

The reality still stuns Rhea. There were thirty-five new astronauts selected. Only six were women.

To the thirty-year-old resident surgeon, the call was probably the most important one she'd receive in her entire life. Rhea had dreamed of becoming an astronaut since she was twelve.

Rhea as cheerleader

Although women were not being accepted into the program, she
was convinced that one day they would be. But it would have to
be an exceptional woman, she decided, one with special training
and abilities. From that point every decision Rhea made about
her life was one that would move her toward that goal. Her first:
to become a physician.

"You do something like that and you'll never get married,"
her peers taunted her.

But Rhea's parents were supportive. Her father had always
stressed self-reliance and reached for high goals. Her mother, who
is now deceased, accepted Rhea's decision with grace, although
she didn't fully understand the reasons.

After high school Rhea left the tree-lined streets of Murfreesboro, Tennessee, the small southern town she had lived in all her
life, and traveled 3,000 miles across the country. There she enrolled in premed at the University of California at Berkeley. The
transition wasn't easy. The enrollment at Berkeley was as large as
the entire town she had left behind. And the free-wheeling
atmosphere was far removed from the all-one-happy-family
feeling she enjoyed in Murfreesboro.

Further, the premed program was academically difficult and
highly competitive. A third of the freshman class didn't make it
to the second year. Rhea struggled but her grades were only
mediocre, nothing like the Honor Society standards she had set in
high school. Her confidence took a tumble.

"I was kind of mixed up, trying to find myself," Rhea says.
"And I knew my grades weren't good enough to get me into med
school."

So she returned to Tennessee and enrolled in a nursing
program at Vanderbilt University. If she couldn't make it as a
doctor, Rhea thought, she could at least be a nurse.

But after a semester at Vanderbilt and another six months at
home, Rhea changed her mind. A nursing degree, although desirable, would not make her eligible for the space program.

"There are times when you just sit and stew," Rhea says.
"Those are the times you grow a little."

The stewing and the growing filled Rhea with determination.
She'd not give up her dream without a fight. She'd go back to
Berkeley and work three times as hard if necessary.

She did. And graduated with honors.

During summers while working as a scrub technician at the Murfreesboro Hospital, Rhea became fascinated with surgery. But surgery is the most exclusive field in all of medicine and would be the least likely to welcome a woman into its ranks. And what about patients? Would anyone want a woman for a surgeon?

"Well, I just made up my mind it would have to be a very special facet of surgery," Rhea says. "So I chose plastic surgery, Let's face it, a man won't go to a female surgeon to have his hernia fixed. But if he were having his sagging jowls lifted or a scar removed, he might consider a woman because she'd be sympathetic."

But after graduation from the Medical School of the University of Tennessee at Memphis and while serving her internship and residency, Rhea became intrigued with a relatively new field of medicine—surgical nutrition. Specifically it's called hyperalimentation. Heretofore surgical patients were fed glucose intravenously, but the body could consume only 600 calories per day. Much more is needed to hasten healing and restore health, particularly with cancer patients. The new system allows concentrated solutions of protein, fat and glucose to be fed intravenously, up to several thousand calories a day. Because the process is innovative and relatively untried, the prospect of experimenting in space was excellent. On Life Science expeditions, for instance, new ways of promoting health will be tried. Rhea decided to specialize.

Meanwhile, she kept in touch with the progress of the space shuttle program at NASA. No longer was NASA concentrating on long-range trips to the moon or Mars. The emphasis of the 1980s would be on a shuttle system that would allow low-altitude orbits for both manned and unmanned exploration. The astronaut of the future did not have to be a test pilot with a string of records to his credit. Mission specialists were needed now, with emphasis on the sciences. Rhea's studies would at least gain her an application.

A pilot's license, although not necessary, would be one more point in her favor, she decided. In the six-month period between her internship and her residency, Rhea worked nights in the emergency rooms of several different hospitals in Mississippi.

Rhea pauses during training exercises. (photo courtesy NASA)

"That left my daylight hours free," she says. "It was spring, the perfect time of the year for flying. So I squeezed in lessons."

She earned her private pilot's license.

In 1976 a "friend of a friend" passed down word that NASA would soon be accepting candidates for the shuttle program, and there was a possibility that women would be included. Training would begin in July 1978.

"Shoot, that's when I'm going to be finished with my surgery," Rhea thought. "It's the perfect time for me." Rhea also knew NASA wanted candidates age twenty-eight to thirty-eight—and she'd be thirty!

"I just addressed a letter to NASA, Houston, Texas," she says. Almost by return mail she received an application.

Rhea applied—along with 8,078 others.

In August 1977 she was summoned to Houston for physical and psychological testing with a group of twenty men and women. On the initial interview she was asked how she'd feel about being cooped up in a space shuttle with mostly men for thirty days or more. How she'd deal with stark living conditions. Would she mind going without shaving her legs? What if they wanted to cut her shoulder-length hair?

"I think my hair looks prettier when it's long, but I'd cut it in a minute if I had to," Rhea says. Stark living conditions and the sacrifices they demanded would be no problem.

As for dealing in close quarters with a group of men, Rhea already had some experience in that area. During internship she was the only woman with a host of male interns. She learned quickly that even her M.D. (Doctor of Medicine) status did not guarantee the same privileges her male cohorts enjoyed. On one of the most trying days she telephoned her father in tears.

"I don't know what to do," she moaned. "The male doctors won't let me wait for my calls in the doctors' lounge because I'm female. The nurses won't let me wait in the nurses' lounge because I'm a doctor. So I've been sitting in the bathroom for the last five hours waiting for my calls. It's terrible."

"There's only one thing to do," her father said. "Go back to the bathroom and wait for your calls, *doctor*."

The bathroom incident was one of the few times in Rhea's life when she has resorted to tears. She's not a crier on earth. She certainly wouldn't be one in space.

One of the tests at NASA involved a rescue sphere, a large "beach ball" made of the same material used for space suits. If Rhea's space shuttle were to be stranded, Rhea would be zipped up in the sphere and floated to a rescue shuttle sent from earth. For testing, she was wired to monitor her heart rate, then closed up inside the ball, only thirty-four inches in diameter. Inside, curled into a fetal position with nothing to see but darkness, nothing to hear but her heartbeat, Rhea lost all sense of time and place.

"They didn't tell us how long it would be," she says, "so if you had a tendency for claustrophobia, it would show up then."

The sphere was just one in an endless array of tests. "In fact we were given so many tests that we began to think everything was a test," she says. "They'd tell us to go get breakfast, and we'd ask if that were a test."

When Rhea returned to the Memphis hospital, she found she had gained a new name. The doctors now called her "Cosmic Rhea."

As she finished the last few months of surgery residency, Rhea tried to push the prospects of becoming an astronaut far back in her mind. After all, NASA would choose only thirty-five, and she was one of more than 8,000 candidates. While she had considered herself prepared and qualified, she found others in her testing group even more advanced. And hers was only one testing group among many.

"The competition was something else," Rhea says. "There were some girls who'd modestly admit to finishing an internship, and later you'd find they had their Ph.D. (Doctor of Philosophy) in chemistry and had done research. I thought, 'Oh my goodness, what have I gotten myself into!' "

In November 1977 Rhea received a call from Abbey asking the same question, "Are you still interested?". At that time he told Rhea the group of candidates had been narrowed to 200—and she was one of the remaining.

When the final call came that January morning, Rhea again confronted skepticism, mixed with a bit of humor. "You must be out of your mind,'' her younger sister Louise, a married school-teacher, gasped. ''You could make $100,000 a year as a surgeon. NASA pays only $22,000.''

Rhea realized, of course, that the sacrifice in income would be considerable, probably greater than that of other astronauts. But her mind was firmly set on the long-range goals. The shuttle program was sure to spur an economic and industrial revolution the way railroads did last century, the way aviation did this century. She envisioned advanced communication systems, the harnessing of solar power, space factories in orbit, space freighters carrying natural resources from the moon and neighboring planets for harvesting on earth.

But more than any of these, Rhea thrilled to the research and manufacture of miracle drugs and vaccines in a weightless, impurity-free environment. One possibility had already been dramatically illustrated aboard the 1976 Apollo-Soyuz space flight. The rare enzyme Urokinase, used for treatment of blood clotting, was separated from human cell cultures six times more efficiently than on earth. The enzyme today costs more than $1,000 a dose, far beyond the reach of the average patient. With the new technology in space, the cost can be cut to $100 or even less.

And there's always the possibility that cancer can be cured in space. Or the common cold.

In July 1978 Rhea set out to make her home in Texas for the next seven years. The NASA-Johnson Space Center is located in Clear Lake City, a small community about twenty-five miles south of Houston. The complex is laid out like a campus, with grassy lawns and a network of paved walkways joining the modern buildings. More than 20,000 people report daily to jobs within the complex, and the area teems with tourists. Only a few miles away, barefoot children go crabbing along the scenic bay shore with only a string and a raw chicken neck for equipment, a far cry from the advanced technology practiced within the confines of NASA.

Rhea's first two years were her candidacy years. NASA was looking the candidates over, she says, and vice versa. ''The selection process must be excellent,'' she adds. ''No one dropped out, and no one was asked to leave.''

Rhea, right, practicing hand washing in zero gravity during parabola in KC-135 (photo courtesy NASA)

Mission specialists such as Rhea must know every system aboard the shuttle—how it works, what happens when it fails, how it interfaces with other systems. The specialists are cross-trained so if one crew member becomes ill, another member can step in and handle the assignment.

During her first year, Rhea received concentrated orientation into the actual NASA operation, the aircraft, astronomy, geology, medicine, craft design, engineering and computers. She learned land and water survival techniques, parachuted through trees and electric wires, lifted weights, ran two miles a day and flew a T-38 training jet.

The second year she began specializing. Because of her background in nutrition, Rhea was assigned to the development of space shuttle food. Then she became involved in payload software, which is essentially the computer programming necessary to support experiments aloft.

Rhea expects to go into space for the first time in the sixteenth shuttle flight. By then the shuttle will be carrying in its cargo bay an actual working laboratory (Spacelab). The Spacelab on the sixteenth flight is scheduled to carry Life Science equipment and experiments. The technology gained on this mission will be carried forward to the next Spacelab, which will be a dedicated Life Science flight.

On this flight the specialists, and perhaps Rhea will be one of them, will be dealing with mysteries yet unsolved. How does the inner ear function in space? Can you gargle? If you spin in a chair, how different is your reaction in space? And what about motion sickness?

The questions may sound simple and unnecessary. But answers to these, and a host of other queries, are needed before the United States can build permanent work platforms and laboratories in space, the ultimate goal of the shuttle program.

The mystery that intrigues Rhea as a doctor is the decalcification of bones that astronauts on previous flights have suffered.

"We know in space there is some muscle wasting and a great loss of calcium in the bones, but we don't know why," Rhea says. "If we find the answer on the shuttle, then maybe we can solve the problem for those on earth, especially old people, who suffer broken bones."

Rhea during parachute training (photo courtesy NASA)

Rhea works with scientists and engineers in designing equipment for use on the shuttle. Because she is the shortest astronaut to date (five feet, three inches) she is used as a model for how high equipment should reach, how far it should span.

Rhea's greatest problem in space will be dealing with zero gravity. Since it can't be duplicated on earth, there's no way to practice. But Rhea and the other astronauts spend hours snorkeling in a water-filled neutral buoyancy chamber, which helps them learn how to maneuver in a weightless condition.

In the sky Rhea has another opportunity to maneuver. The KC-135 military aircraft climbs and dives so steeply that it produces thirty seconds of zero gravity at the top of each parabola. During this half minute, repeated over and over each flight, the astronauts grab the opportunity to experience weightlessness. Moving from zero gravity to pulling two G's (twice the force of gravity) at the bottom of the parabola in a matter of seconds often

Rhea during orientation briefing (photo courtesy NASA)

causes motion sickness. The astronauts call the plane their "Vomit Comet."

It was in the "Vomit Comet" that Rhea experienced her most exciting project to date—designing equipment that could save a life aboard the shuttle. Using a life-size mannequin, Rhea and the other specialists aboard tried again and again to perform cardio-pulmonary resuscitation and mouth-to-mouth breathing while the plane roller-coastered. But it was exasperating.

"It's such a different thing in zero gravity," Rhea says. "The mannequin floats away, then you float away, and you don't have enough weight to compress the chest. You can't lean on it like you do here on earth because you don't weigh anything."

Fighting back nausea, they tried again at the top of the parabola. Again they failed.

"After about forty parabolas we just said, 'Look, let's start back at the beginning and see how we would have designed this equipment if we had started out with nothing,'" Rhea says.

It worked. The specialists designed a restraint system that allows the rescuers to push up with their legs, using an entirely different set of muscles.

"It's a whole new approach. We were kind of proud of it," Rhea says. "It wouldn't work in an earth environment, however."

Rhea must deal with the prospects of being the only medical doctor aboard a shuttle. If there were an emergency appendicitis, would she operate? Certainly, she says. Although she wouldn't know what to expect. Would blood flow or gush in zero gravity? Would it coagulate? What if the instruments floated away? There's no way of knowing.

In most emergencies there would be time to bring the shuttle back to earth. But Rhea must be prepared for a trauma aboard— an accident or injury requiring immediate treatment.

Rhea and the other astronauts will leave the packing for their space trips up to the government. They'll be provided a bag of government-issued items, mostly "off-the-shelf" products to prevent the appearance of endorsing certain manufacturers. The clothing will be primarily unisex intravehicular garments. A pair of pants, a jacket and shirt, plus a set of underwear and something to sleep in.

When the women were shown their choice of underwear—boxer or jockey shorts—they, of course, requested something else. Now they can choose between bikini or regular panties, plus a bra to their liking.

All clothing will be cotton to avoid sparks sometimes caused by synthetics. For sleepwear, the women selected shorts and a T-shirt, because they will double as leisure clothes in off-duty hours.

Previously the government issued personal hygiene kits brimming with toothpaste and brush, shaving cream and man-type deodorants. Now the women's kits will carry lipstick, hand cream and tampons.

The atmosphere aloft will be shirt sleeve, with the temperature ranging from a comfortable 70 to 75 degrees Fahrenheit and the oxygen content duplicating that on earth. The astronauts will not be required to wear helmets or space suits, not even for lift-off and reentry. The living cabin will be pressurized like an airplane. If it's necessary to leave the cabin, the astronaut will go into an air lock, put on a special space suit, then depressurize the air lock, just as the men who walked on the moon did. The living quarters will remain pressurized.

As far as area is concerned, living in a shuttle will be like setting up residence in a large-size bathroom or a small cubicle office. "And for seven people per trip that could get pretty crowded," Rhea says. The astronauts have worked closely together, however, and have learned to interact with one another and to avoid friction.

The shuttle is equipped with a small lavatory with a privacy curtain, and a toilet with a door. Each bunk bed has a door, a reading lamp and taped music for relaxation.

As for dealing with a menstrual period in space, Rhea says they, of course, do not know what to expect. But they're not making a big deal of it. "At the moment we just proceed as if there are no problems to be anticipated on that score," she says. Later NASA may schedule a shuttle trip to coincide with an astronaut's period to purposely test the reaction.

The astronauts will work a sixteen-hour day while in orbit. How much time they'll have to sit and relax Rhea doesn't know. But she plans to take advantage of every minute. Instead of curling up with a book, which she can do at home, she'll watch the scenery.

"At 150 miles up you get a tremendous overview of the earth," she says. She'll use a camera and binoculars supplied by NASA. She's had courses in geology and geography, plus a course in star recognition. And she took a field trip to New Mexico to examine faults, where earth plates are slipping across one another or have broken in particular ways. In her study she used photos taken from Skylab, an indication of the importance of examining the earth from above.

Rhea (first on left) with five other astronauts and their rescue ball. The others are, left to right, Kathryn Sullivan, Judith Resnik, Sally Ride, Anna Fisher, Shannon Lucid. Behind Rhea is a mannequin in extravehicular suit. (photo courtesy NASA)

As an astronaut, Rhea lives what she calls "an ordinary life-style." She has a car and an apartment with modern furnishings. She cooks and cleans house, shops at the supermarket, grows a garden with spinach, tomatoes and snow peas. She even finds time to nurture and weed a flower garden with zinnias and marigolds.

All of this is coordinated into a sixty-hour work week at NASA, plus forty-eight hours of moonlighting every month as an emergency room physician at a nearby hospital. Because she is not actually dealing within her own profession in the shuttle program, NASA makes it possible for her to moonlight in order to maintain technical proficiency. NASA also sends her to several medical conventions and seminars each year.

Until recently Rhea lived alone. But on May 30, 1981, in her home town of Murfreesboro, she became the bride of Robert Lee Gibson, an astronaut and former Navy fighter pilot. They met in training, became the first two astronauts to marry after entering the program, and thus dispelled the fears of her girl friends back home that she'd be doomed to spinsterhood forever.

After her seven-year tenure is completed, Rhea may elect to return to medical school for a Ph.D. (Doctor of Philosophy) in surgical nutrition, if the field is not yet too crowded. Or she may yet specialize in plastic surgery. But the decision will be a difficult one. Her training at NASA is opening many new fields in computers, machines and aviation. "There are so many neat things to do in life," Rhea says. "I'd like to try lots of them."

She may even decide to remain with NASA.

"Being with NASA isn't like being a professional athlete with a contract coming up for renewal each year," Rhea says. "You just stay with NASA as long as you're happy with them and they're happy with you."

Rhea's decision to become an astronaut wasn't one she could take lightly. There are inherent dangers involved that could affect her entire life. The biggest is probably the exposure to radiation. Now that she is married, she would like to have children. Would radiation affect her ability to have healthy children? To carry a pregnancy full term?

Although she would not want to give birth in space, she would not hesitate to become pregnant during her program.

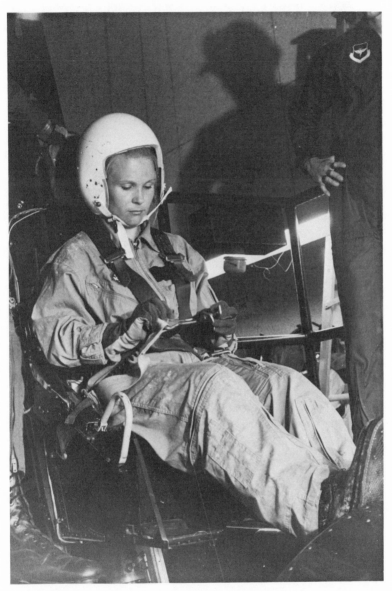

Rhea buckles her seat belt. She will shortly be "ejected" in a simulated parachute drop. (photo courtesy NASA)

"We're all the kind of women who would work right up to the time of delivery and go back to work right away," Rhea says. "As long as we handle it intelligently, there's no reason why we couldn't have babies during the program."

To Rhea the space program is the frontier for her generation. There'll be others—the sea, the interior of the earth—and Rhea may be part of them, too.

But at the moment her full dedication is to the shuttle program and its success. She maintains an enduring faith in science and technology, in their ability to work out any problems they may encounter. While she doesn't dwell on the dangers, she realizes they are ever present.

"We'll just do the best we can at all times," she says. "And if it kills us, then that's that. It's the risk we take to be an astronaut."

JUDY ANN LEE

☐ FLIGHT ENGINEER ☐

JUDY PEERED THROUGH the darkness outside the cockpit window, her gaze riveted on the string of lights along the shore, the sparkle of spacious Miami Beach hotels in the distance. As the sleek Grumman Gulfstream II jet nosed toward the Fort Lauderdale, Florida, airport, something inside Judy clicked, the personal sensor that signals "Take note, this is for you."

"It was the most beautiful sight I had ever seen," Judy says. "I knew that instant that aviation was for me, that one way or another I'd work my way to the cockpit again."

But Judy meant for her next position in the cockpit to be as a crew member rather than as a curious spectator. She had no way of knowing the time and effort, the sacrifices and patience necessary before her newfound goal could be realized. Judy was eighteen and a college freshman, stirred by personal ambition and an excitement for life. But the year was 1970, and the cockpits of commercial aviation in this country were still firmly locked against the keys of women.

Judy's flight in the Grumman was a happenstance, a serendipitous ride home from the wedding of her best friend in South Carolina. Since it wasn't an official business trip, the crew invited Judy to the cockpit, her first visit up front.

"I was amazed at the complexity of flying, the teamwork of the crew members as they worked the plane in for a landing," she recalls.

The next day Judy went to the Fort Lauderdale airport, plunked a $10 bill on the counter and went for a twenty-minute

Judy and the plane she flies

spin in a Cessna 150, a small two-seater with a one hundred-horsepower engine. To Judy the Cessna looked like a toy compared with the business jet from the night before. But the instructor explained a few basics of flying, told her about ground school, about building time in the air to progress from one plane to another. Judy fingered the rudder, the instrument panel. She tried on the headset. The thrill was still there. The sensor inside was still sending positive signals. Judy signed up for lessons.

Now Judy is part of a three-member crew aboard a 727 passenger jet, the plane she affectionately refers to as the "meat and potato" of the airlines. Now she knows firsthand the intricacies of the cockpit, as well as the joys. The time span from that first trial spin to signing on with United Airlines as flight engineer, the fifth woman pilot to be hired by United, took slightly less than eight years. It could well have taken longer. With most women it does. But Judy modestly contends she was "in the right place at the right time." Even so, she is still not at the controls, actually flying the plane. That reality lies some distance in the future.

While Judy's modesty is appealing, it isn't necessarily valid. If she had not been highly qualified, the job would not be hers. To gain those qualifications required "tunnel vision," a term Judy uses to describe the act of keeping her eye firmly on her goal, not giving way to setbacks and discouragement, and biding her time until cockpit doors would swing open to women. All these things added up to one total—a big gamble.

Yet Judy took the dare.

She attended ground school two nights a week and skipped college classes to take flying lessons during the day. After a flight she'd often linger at the airport, mesmerized by the accomplished landings and takeoffs of the thundering passenger jets. "Just don't even think of it, Judy," her instructor advised one day as he found her at the edge of the runway. "They'll never hire a woman."

A few years later, after Judy had earned her private license and was slowly building up flying time, a Navy enlistment officer visited her school, Florida Atlantic University, to speak on career opportunities in the military. Judy, enthusiastic about an opportunity she hadn't considered, asked about the possibilities of flying in the Navy. The officer laughed. "The uniforms won't fit you," he said.

So Judy learned to share her ambition with only her parents and a few close friends. She also learned that to continue in aviation would require more and more money. Reluctant to borrow from her parents, she discovered funds from an unusual source.

Her horse.

Judy had ridden since she was three. At ten she made a deal with her father. If she saved enough money to buy her own horse, he'd pay for the upkeep. It took Judy four years, during which time she banked every penny that passed through her fingers— fees she earned doing odd jobs, allowances, birthday and Christmas gifts from uncles and grandparents. She bought Kilkenny at a riding stable for a ridiculously low price for such a fine mount, a part thoroughbred, part quarter horse. Then she and Kilkenny signed with a trainer and went through rigorous schooling. By the time Judy entered aviation she had become an accomplished equestrian.

Kilkenny proved a money winner in shows from Florida to New Jersey. With Judy in the English saddle, he captured purse after purse. Judy banked the winnings in a special ''horse'' fund with one purpose in mind—to retire Kilkenny in style when he grew too old to jump and show. Occasionally Judy dipped into Kilkenny's fund to finance another series of flying lessons, but it was always just a loan.

''I invariably paid it back,'' she says. ''Every penny of it. The money was Kilkenny's, not mine.''

But Kilkenny gave his owner something that required no paying back, a gift of lasting value. In a strange way, the years of intense training with her horse developed in Judy the same qualities she'd need to become a pilot—persistence, self-discipline, study, a willingness to sacrifice and lots of patience.

Judy takes a hurdle with Kilkenny, 1970. (photo by Hank Cohen)

Judy, like all horse lovers, shared a special relationship with Kilkenny, a mutual respect, a mutual goal. Leaning forward to take a jump, her hands loosening the reins to give Kilkenny his head, Judy felt as though the horse's body became an extension of her own, as though the same thought raced through the two of them. Now she has the same feeling about an airplane. It responds to her touch, obeys her bidding.

After Judy graduated from Florida Atlantic University with a degree in European history, she signed on with Hialeah Race Track as exercise girl for some of the world's fastest racehorses. The work was dangerous. Spills were frequent and could result in broken bones, internal injuries or permanent damage. The work was also strenuous and, like aviation, demanded a good degree of physical fitness. Exercisers are usually boys or small men with immense strength. But race track owners are now finding that young women make excellent exercisers if they build the right muscles. And again like aviation, women become involved for a special reason—the sheer love of it.

For Judy the job had another appeal. The horses are exercised in the mornings. This gave her free time in the afternoons for flying.

Judy traveled with the Hialeah stables in the Northeast and Midwest. Then when their seasons were over, she joined other stables and set out for different circuits. Leaving her own horse behind, she packed her jeans and T-shirts, her stereo, a small television set and her white German shepherd dog, Killer, into her compact car and took to the road.

"I think I was the sole support of the Holiday Inn chain during those years," Judy says. "They always let Killer stay in the room with me at no extra charge. My dog was my best friend. With her I always felt safe, and my parents didn't worry so much."

At the track she left Killer in the tack room while she worked, or tied her to the barn to await Judy's return.

Judy ended up in California, where she signed on with Charles Whittingham at the popular Santa Anita Race Track. Whittingham, one of the world's most famous and successful trainers, viewed Judy with skepticism when she applied.

"A little thing like you can't gallop a racehorse," he teased.

Judy is five feet, five inches tall. At the time she tipped the scales at a mere one hundred pounds.

"I can," she said. "I can. I really can!"

She could. And she got the job.

Exercising began at 5:30 or 6 A.M. and involved taking the horses to the track and putting them through whatever paces the trainer ordered. Some days Judy galloped as many as thirteen horses before 10 o'clock. Other mornings she concentrated on one horse, calming a nervous colt, familiarizing it with the track. Her most demanding task was "breezing" a racehorse by letting him run through the gate and build up speed on the track as a sort of "dress rehearsal" for the afternoon's race.

"I didn't know how long I'd last on that job," Judy says. "It takes a lot out of you physically, and the pay is certainly no incentive."

Judy earned about $150 a week, plus a bonus if a horse she trained won a special purse. Out of this she had to pay her own room and board, and her own travel expenses when the stable moved from track to track. Judy spent nothing on clothes or cosmetics and didn't even own a dress. Thus she was able to reserve enough money to continue flight training at a nearby airport. She gained her commercial pilot's license. Her hours in the sky were mounting.

Even more important, she began hearing rumors that major airlines were beginning to hire pilots again after a severe cutback caused by the fuel embargo of 1973. And this time there was a slim chance a few of them would be women.

Judy mailed applications to all the large airlines and set up a card file of their responses. One company repeatedly mailed her an application for stewardess training, although she had clearly listed her qualifications for pilot. It took a tersely worded letter to the company's president to pry loose a pilot's application for Judy. All the other lines zipped back to her the same message, "Not hiring now but update your application every year."

Judy did more than that. She updated each application every three or four months. Meanwhile, she borrowed money from her best girl friend (the "horse" fund was now being spent to retire Kilkenny), returned to her parents' home in Fort Lauderdale and began serious study for her instructor's license. If it took an impressive number of hours in the sky to gain the attention of an airline, then instructing was the fastest way to build those hours.

"My friends couldn't believe what had happened to me," Judy says. "I've never been much of a student and they knew it. Then I suddenly started spending every waking hour with my nose in an aviation book. The system changes so rapidly that you have to constantly be in touch to keep up to date. I had fallen behind in California."

She soon made it up, got her instructor's license, then took on an instructor's job at an airport near her home, working every day as long as there was daylight.

"It was even harder than exercising horses," Judy says. "I spent all my time at the airport either waiting for students or flying any odd jobs I could get. I flew air taxi, charters and made blood runs (flying blood plasma to hospitals in emergencies)."

Living at home, she used every penny she earned to pay back her friend. Her air time was climbing, her application file bulging.

Then in late 1977 United Airlines called to say they were interested. Judy's first interview was in Atlanta, Georgia, for which she paid her own expenses. For the next tests in Washington, D.C., she flew free but paid her own expenses while there. But from that point on, United picked up the full tab. Judy underwent exhaustive testing for mechanical ability, psychological aptitudes, personality profiles and general knowledge, plus a thorough physical. In Denver, Colorado, she was trained and tested in a simulator.

On the final interview Judy was asked what she planned to do if she didn't get the job. "I just don't think in those terms," she replied. "I plan to get the job."

She did.

Judy finds airlines operate in an extremely democratic manner, using seniority as the basis for promotions, job headquarters and choice of schedules. She, like all the other pilots, began as a second officer, or flight engineer, even though she is a fully qualified pilot. The copilot serves as first officer and the pilot as captain. It will probably be five years before Judy can move forward to the copilot's seat and, along with the pilot, actually fly the airplane. As for when she'll have ample seniority

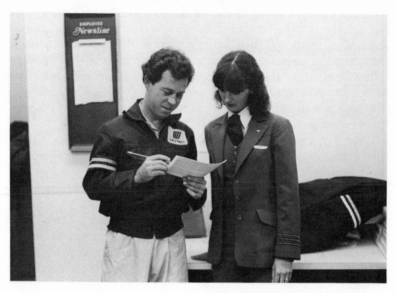

Judy checking fueler's report before flight (photo by Elizabeth Smith)

to become captain, Judy says only, "Who knows? Ten years? Fifteen? It's only a guess."

As flight engineer, Judy is in charge of the preflight condition of the plane. She checks the exterior, the interior, the tires and the basic airframe. She tests the oxygen masks over the seats, the heating and cooling systems inside the plane. She prepares the preflight paper work and files the takeoff and landing cards.

While in flight, Judy sits at a panel behind the copilot and monitors the systems to be sure the fuel is balanced, the engine not overheating, the inside temperature at a comfortable level.

Judy has already accumulated enough seniority to bid for certain locations. She exercised this recently when she chose Los Angeles, California, as home base. She shares a house with another woman in a small suburban town with quaint shops and sidewalks suitable for roller skating.

Judy flies about 77¹/₂ hours per month, plus layovers, which means she's away from home about 50 percent of the time. The knowledge that her schedule would pose a problem to a married pilot doesn't bother Judy. If she marries it will be to someone

Judy at instrument panel (photo by Elizabeth Smith)

who understands. The dating game to her is phony. She enjoys campanionship with small groups, going out for dinner or to musicals. But large, loud parties would never appear on her list of things she'd like to do. She'd rather stay home and read a book or watch a movie on television.

At United the crew flies together for a thirty-day period before changing schedules. During this period, they share a camaraderie and usually become good friends, although they may not fly together again for years. Only once has Judy flown with a captain who obviously didn't want a woman in his cockpit.

"He completely ignored me in his conversations," Judy says. "And when he wanted me to do something, he very rudely ordered me rather than asking in a nice way."

The condition persisted the entire thirty days. But every woman pilot runs into a captain like that at least once in her career, Judy says. Fortunately such captains are in a minority now. "It bothered me a great deal," Judy says, "but I tried not to show it."

At Thanksgiving and Christmas Judy bids for flights in order to relieve married pilots so they can be with their families. "We always make a special deal of holidays when we're away from home," she says. "You know, homemade cookies, special dinners, that sort of thing."

Judy enjoys the travel, the change of scenery any time of the year. For her the sheer beauty of Portland, Oregon; the excitement of San Francisco, California; or merely staying at the enchanting old Palmer House in Chicago, Illinois, are all special bonuses. But none of them compares with the joy of flying into a sunset!

Airline pilots have a saying they frequently quote about their jobs: "It's so great you have to work at making it miserable." Judy agrees.

Another good aspect of her job is the income. "I don't spend money frivolously," she says, "but I like a nice lifestyle. I'm not saying I want diamonds and furs. But if I'm walking down the street and see a special sweater in a shop window, I want to feel I can buy it."

Judy has driven her compact car for six years. This year she's trading it in for a new BMW. "Lots of pilots drive more expensive

Judy with two other members of ISA in Denver, Colorado, 1980

cars, a Porsche or a Jaguar or a Mercedes,'' she says, ''so I don't think I'm going overboard with a BMW.''

Meeting pilots from other airlines rates high with Judy. She serves as secretary for the International Social Affiliation of Women Airline Pilots (called ISA for short), a social organization for ''getting together and getting acquainted.'' This year its 140 members from all over the free world met in Cancun, Mexico.

Judy still has an abiding love for horses and often goes to the race track to see her old friends. But she's never tempted to climb into the saddle and gallop a racehorse full speed.

''When you love flying the way I do, you quit doing dangerous things and concentrate on good health,'' she says. She, like all pilots, maintains a regular program of exercise. For Judy it's jogging, aerobic dancing, bicycling and riding a pleasure horse. Every year she must pass a stringent physical examination. But if she stays in shape she'll be able to fly until age sixty, the mandatory retirement age at United.

And she intends doing just that.

A SELECTED CHRONOLOGY OF WOMEN IN AVIATION

1784   Mme. Tible becomes first woman to ascend in a balloon. She was not the pilot. (June 4, France)

1819   Mme. Marie M. S. Blanchard becomes first woman to die in an air tragedy. The balloon in which she was riding ignited from fireworks attached to the rear. (France)

1863   Mlle. Eliza Garnerin becomes first woman to make parachute drops. (France)

1880   Mary H. Myers becomes first American woman to pilot her own balloon.

1898   Mrs. D. Klumpke-Roberts becomes first astronomer to leave earth for the purpose of studying stars from a balloon. (San Francisco, Calif.)

1903   Aida de Acosta becomes first woman to pilot a powered aircraft solo. She was an American. (Paris, France)

1908   Mlle. P. van Pottelsberghe and Mme. Thérèse Peltier both claim to be the first woman to fly as passenger in an airplane. Pottelsberghe flew in Belgium and Peltier in France.

1909   Mlle. Marie Marvingt becomes first person, man or woman, to make a balloon flight across the North Sea.

Baroness Raymonde de la Roche makes first airplane flight by woman pilot, piloting a Voisin 350 yards. (October 22, France)

1910   Baroness de la Roche becomes first woman to earn a pilot's license. (March, France)

Mlle. Helen Dutrieu becomes first woman to fly in an aviation meet, although there were no prizes or awards for women. Mlle. Dutrieu was French. (Spring, Belmont Park, N.Y.)

Blanche Scott becomes first American woman to solo, flying from one end of a driving park to the other. (Fort Wayne, Ind.)

1911   Harriet Quimby becomes first American woman to earn a pilot's license. (August 1)

1912   Harriet Quimby becomes the first woman to solo across the English Channel. (April 16)

Georgia ("Tiny") Broadwirck becomes first woman to parachute from an airplane. (Los Angeles, Calif.)

Lillian Todd becomes first woman to design her own aircraft. (New York, N.Y.)

1914   Mrs. L. A. Whitney becomes first woman passenger on a regularly scheduled airline. (January 8, St. Petersburg to Tampa, Fla.)

Emma Stinson becomes first woman to operate a flying school. Her son became her partner. (San Antonio, Tex.)

1916   Katherine Stinson becomes first night-flying woman sky writer. She used flares. (September, Los Angeles, Calif.)

1918   Katherine Stinson becomes first woman to become member of the air mail staff. (September 26, Washington, D.C.)

1919   Sylvia Boyden becomes first professional woman parachutist in the United States.

1922   Lillian Gatlin, as a passenger in a mail plane, becomes first woman to make coast-to-coast flight. (October 8)

1925   Mrs. Irene MacFarland becomes first woman member of the Caterpillar Club, an unofficial organization of people who have been forced to bail out of an airplane in an emergency and survived. (July 4)

Ruth Gillette becomes first woman to enter a national aviation event. (September 27, Long Island, New York)

1928   Marjory Stinson becomes first woman to fly as passenger over the Panama Canal Zone, coast to coast. She gained special permission from the U.S. Army. (January 24)

Lady Mary Heath, as second pilot with Royal Dutch Airlines, becomes first woman pilot in air passenger service. She did not fly regularly. (July 28, Amsterdam, Holland, to London, England)

1929   The first Women's Air Derby is held. Louise Thaden won in heavy aircraft, Phoebe Omlie in light aircraft. (August 18-26, Santa Monica, Calif., to Cleveland, Ohio)

Ninety-Nines, first organization of women pilots, is organized. (November)

1930   Laurette Schimmoler becomes first woman to manage an airport. (Port Bucyrus, Ohio)

United Air Lines hires the first hostess, Ellen Church. (May 15)

1932   Amelia Earhart becomes first woman to solo across the Atlantic, flying from Harbour Grace, Newfoundland, to Londonderry, Ireland. (May 20-21)

Amelia Earhart completes first nonstop transcontinental flight by a woman, flying from Los Angeles, Calif., to Newark, N.J. (August 24-25)

1933   Amy Johnson Mollison becomes first woman to cross the Atlantic westward. (July 23)

First issue of *The 99ers* is published, the first magazine devoted entirely to women in the air. (November)

1934   First women's National Air Meet is held. (August 4-5, Dayton, Ohio)

Laura Ingalls becomes first woman flier to circuit South America.

Phoebe Omlie is appointed special assistant for Air Intelligence, the first airwoman to become a federal government official. (December)

1935   Phoebe Omlie, Helen Richey, Helen McCloskey, Louise Thaden, Nancy Harkness and Blanche Noyes begin the first air-marking campaign.

Amelia Earhart becomes the first woman to participate in the Bendix Race from Los Angeles, Calif., to Cleveland, Ohio. She finished fifth. (August 30)

Jean Batten becomes first woman to solo the South Atlantic, flying from England to Brazil. (November 15)

1936   Louise Thaden becomes the first woman to win the Bendix Trophy Race. (September 4, New York, N.Y., to Los Angeles, Calif.)

Jean Batten becomes the first woman to fly England to New Zealand. (October)

1937   Amelia Earhart and Fred Noonan start around the world from California to New Guinea and are lost at sea in the South Pacific. (May 21-July 2)

1938   Jacqueline Cochran wins Bendix Trophy Race. (September 3, Burbank, Calif., to Cleveland, Ohio)

1939   Jacqueline Cochran makes first blind landing by a woman, using an instrument landing system. (August 8, Pittsburgh, Pa.)

Women are admitted to the Civil Aeronautics Administration Pilot Training Program on the basis of one woman for each ten men.

1941   Jacqueline Cochran copilots an Air Force bomber to England. (June 19)

Stephen's College, Columbia, Mo., becomes the first school to offer women a course in aviation. (September)

Women are eliminated from the Civil Aeronautics Administration Pilot Training Program after 2,250 have been trained. They were replaced by men training for war.

1942   The Women's Auxiliary Ferry Squadron (WAFS) and the Women's Airforce Service Pilots (WASPs) are formed to aid the war effort. Its members are promised military status. (September)

Women Accepted for Voluntary Emergency Service (WAVES) begin serving in naval aviation in jobs other than piloting.

1943   The first women are trained as air traffic controllers for the Civil Aviation Administration (forerunner of the Federal Aviation Administration) due to a shortage of men. After the war they were replaced by men, and it wasn't until several decades later that women were again admitted to the control tower.

1944   The WASPs are abruptly disbanded without ever having been granted military status. (December 20)

1950s The Soviet Union hires its first female commercial pilot. (Exact year unknown)

1953 Jacqueline Cochran becomes first woman to break the sound barrier. She was granted the privilege of flying an F-86 Sabrejet by the U.S. Air Force.

1961 Turi Wideroe becomes first woman commercial pilot for a major airline in the free world when she signs on with Scandinavian Airlines.

1963 Valentina Terechkova, Russian cosmonaut, becomes first woman in space, piloting a spacecraft through forty-eight orbits.

1966 Ensign Gale Ann Gordon becomes first woman to solo in a Navy training plane. (March 25, Saufley Field, Pensacola, Fla.)

1970 Girls are admitted to Junior Air Force Reserve Officer Training Corps (ROTC) for pilot training in high school. (Autumn)

Women are admitted to Air Force Reserve Officer Training Corps (ROTC) for pilot training in college. (Autumn)

The first women are admitted to U.S. Air Force Officers Training School. (Autumn)

1971 Ruth M. Dennis becomes the first female chief of an air service station for the Federal Aviation Administration. (April 4, San Diego, Calif.)

Gene D. Sims becomes the Federal Aviation Administration's first female tower chief. (May 16, Cuyahoga County Airport, Ohio)

1973 The first class of eight women begins pilot training with the U.S. Navy. (March 2)

Lieutenants Victoria M. Voge and Jane D. McWilliams become first female flight surgeons with U.S. Navy. (December 20)

Lieutenant Junior Grade Shelly Robinson becomes first woman admitted to Air Traffic Control Officers School in the U.S. Navy. (June, Glynco Naval Air Station, Illinois)

Emily Warner, as copilot with Frontier Airlines, becomes first American woman to fly for a major passenger airline. (January)

1974   Lieutenant Barbara Ann Allen (now Rainey) becomes first woman to receive Navy Wings. (February 2)

1976   The first women are admitted to the U.S. Air Force Academy. (June)

The first women are admitted to the U.S. Navy Aviation Officer Candidate School at Pensacola, Fla. (June)

The first women are admitted to pilot training in the U.S. Air Force. (August 23)

Emily Warner becomes first female captain in American aviation while flying for Frontier Airlines. (June 1)

Lieutenant Sharon McCue becomes first woman to serve as aviation maintenance duty officer with U.S. Navy. (June)

1977   The first women are admitted to navigator training in the U.S. Air Force. (March 27)

WASPs are declared veterans of World War II and given honorable discharges thirty-three years after they were disbanded. (November)

Lieutenant Janna Lambine becomes first female pilot for the U.S. Coast Guard when she flies a helicopter. (March 4, Whiting Field, Milton, Fla.)

1979   The National Aeronautics and Space Administration (NASA) admits the first six women for training in the space shuttle program. (July)

1980   The U.S. Air Force Academy graduates its first class to include women.

GLOSSARY

aerobatics—spectacular flying feats, such as rolls, dives, etc.

aerodynamics—relating to motion of air and bodies in the air.

air lock—an air chamber between the outer atmosphere and the cabin.

alert facility—fenced area and building for airplane and personnel "on alert," or in constant state of readiness.

authenticator—computerized device for verifying a message or mission as an added security measure.

balloon—bag made of lightweight but durable material filled with gas or heated air to make it rise and float.

boom—a movable arm.

boom pod—small area in rear of airplane from which boom is operated.

Caterpillar Club—an unofficial organization of people who have been forced to bail out of an airplane in an emergency and survived.

charter—to rent a plane and its crew for a special trip, or a plane that has been rented.

Civil Liberties—an abbreviated term for the American Civil Liberties Union, an independent group formed in 1920 to champion the rights of people as set forth in the Constitution.

classified information—information not intended for the general public.

configure the plane—to plan distribution of weight in relation to the center of gravity through the use of charts and graphs.

detask—to relieve from an assigned mission.

dirigible—a lighter-than-air aircraft with a means for propelling and steering.

down to minimums—indicates the pilot is relying on instrument readings rather than her own vision due to bad weather and is landing under rigid restrictions.

earth plates—movable outer surfaces of earth. Volcanoes and earthquakes occur where the plates meet.

enlisted—a member of the armed forces holding a rank below Second Lieutenant or Ensign.

extravehicular—for use in space outside the shuttle.

fault—a fracture in the earth's crust.

G—a unit of force applied at rest equal to the force exerted by gravity.

general aviation—nonmilitary and nonpublic aviation, such as planes privately owned or owned by corporations.

Gold Star—a symbol, usually appearing in a front window, denoting a soldier from that home was killed in war.

ground school—a school giving lessons in aerodynamics, map making, photography and other subjects for aviators.

interceptor—a light, high-speed, fast-climbing military plane.

intravehicular—for use within the space ship.

ISA—abbreviation for International Social Affiliation of Women Airline Pilots, a social organization.

klaxon—a shrill, electrically operated warning signal.

mayday—an international radiotelephone distress signal.

Ninety-Nines—the first organization of women pilots.

offload—the amount of fuel specified to be delivered to the receiving plane.

parabola—the arc formed by the ascent and descent of a plane.

para-sail—a parachute pulled by rope behind a boat until proper altitude is reached.

para-sailing—parachuting into water through the use of para-sails.

Relief Wings—an organization of volunteer pilots formed early this century to fly emergency relief to stricken areas.

scramble—summon for quick takeoff.

secure the cargo—tie it so it does not roll around.

secure the hatches—close and lock securely.

sextant—an instrument for measuring altitudes of celestial bodies from a moving airplane.

simulator—airplane controls set up in a laboratory to duplicate the actual act of flying.

slipstick—a slide rule.

slip-way doors—unhinged doors that slide open under pressure.

sortie—two or more airplanes on an assignment.

Strategic Air Command (SAC)—a branch of the Air Force that responds to nuclear attack.

tactical—a branch of the Air Force that remains in a state of constant readiness to assist ground warfare.

tail wind—a wind having the same direction as the airplane.

TRACON—abbreviation for Terminal Radar Control. The room in which the radar control is housed within an air traffic control tower is called the TRACON room.

WASP—abbreviation for Women's Airforce Service Pilots, an organization whose members served during World War II.

WAVES—abbreviation for Women Accepted for Volunteer Emergency Service, an organization whose members served the U.S. Navy during World War II.

wing-walk—to actually walk on wings while plane is in flight.

wings—an insignia shaped like outspread bird wings awarded on completion of prescribed training. Wings are usually worn on the breast of a uniform.